New Directions in the Philosophy of Science

Series Editor
Steven French
Department of Philosophy
University of Leeds
Leeds, UK

The philosophy of science is going through exciting times. New and productive relationships are being sought with the history of science. Illuminating and innovative comparisons are being developed between the philosophy of science and the philosophy of art. The role of mathematics in science is being opened up to renewed scrutiny in the light of original case studies. The philosophies of particular sciences are both drawing on and feeding into new work in metaphysics and the relationships between science, metaphysics and the philosophy of science in general are being re-examined and reconfigured. The intention behind this new series from Palgrave Macmillan is to offer a new, dedicated, publishing forum for the kind of exciting new work in the philosophy of science that embraces novel directions and fresh perspectives. To this end, our aim is to publish books that address issues in the philosophy of science in the light of these new developments, including those that attempt to initiate a dialogue between various perspectives, offer constructive and insightful critiques, or bring new areas of science under philosophical scrutiny.

The members of the editorial board of this series are: Holly Andersen, Philosophy, Simon Fraser University (Canada) Otavio Bueno, Philosophy, University of Miami (USA) Anjan Chakravartty, University of Notre Dame (USA) Steven French, Philosophy, University of Leeds (UK) series editor Roman Frigg, Philosophy, LSE (UK) James Ladyman, Philosophy, University of Bristol (UK) Michela Massimi, Science and Technology Studies, UCL (UK) Sandra Mitchell, History and Philosophy of Science, University of Pittsburgh (USA) Lydia Patton, Philosophy, Virginia Tech (USA) Stathis Psillos, Philosophy and History of Science, University of Athens (Greece)

Initial proposals (of no more than 1000 words) can be sent to Steven French at s.r.d.french@leeds.ac.uk.

More information about this series at
http://www.palgrave.com/gp/series/14743

Rolf Hvidtfeldt

The Structure of Interdisciplinary Science

palgrave
macmillan

Rolf Hvidtfeldt
Aalborg University
Copenhagen, Denmark

New Directions in the Philosophy of Science
ISBN 978-3-030-08119-5   ISBN 978-3-319-90872-4   (eBook)
https://doi.org/10.1007/978-3-319-90872-4

Cover illustration: John Lund

Printed on acid-free paper

This Palgrave Macmillan imprint is published by the registered company Springer International Publishing AG part of Springer Nature.
The registered company address is: Gewerbestrasse 11, 6330 Cham, Switzerland

Series Editor's Foreword

The aim of this series is to map out exciting new directions in the philosophy of science. Sometimes, this leads to familiar ground being viewed from a fresh perspective; at other times we are taken somewhere entirely different. Such is the case here. As Rolf Hvidtfeldt notes in his introduction, 'interdiscplinarity' is very much a 'hot topic' these days, yet it has not been philosophically analysed. Deploying the tools of philosophy of science—in particular those developed through consideration of scientific representation—Hvidtfeldt offers a novel framework for evaluating the *epistemic* virtues and vices of interdisciplinarity. This stands in stark contrast to the work one tends to find in 'Interdisciplinarity Studies' where a focus on social factors dominates.

Taking representation to be the central activity of science, Hvidtfeldt insists that the effects of interdisciplinarity must be reflected in the relevant representational activities as manifested in scientific publications. In particular, he focuses on the notion of a 'scientific approach', in the sense of a '*a specific way of using a specific vehicle to represent a specific target.*' Identifying the representational vehicle and target may be more or less straightforward, but between the two lies a 'conceptual layer' of assumptions, tools and models, both physical and mathematical, that is often overlooked. This approach-based framework can then be used to supply an analysis of interdisciplinarity in which existing vehicles of representation can be understood as imported into a new domain or hybridised in interesting ways, with consequent changes to that mediating 'conceptual layer'.

As an illustrative example, Hvidtfeldt looks at a specific interdisciplinary project within research into schizophrenia. Drawing on his own training in the fundamentals of this project, he is then able to identify the relevant vehicles of representation and the associated targets, as well as the various elements of the conceptual layer that are combined, modified and, in some cases, distorted almost beyond recognition. Contentiously perhaps, Hvidtfeldt argues that although this particular project may enrich our understanding of schizophrenia, his application of the approach-based framework reveals certain crucial problems that result from not paying sufficient attention to the difficulties involved in bringing together distinct approaches within an interdisciplinary context. As he concludes, the non-universal nature of certain elements of representation means that importing them into a new context runs the risk of violating the relevant ceteris paribus clauses. And when the scientists involved do not possess the 'deep' expertise associated with the field in which those elements were first developed, serious problems may then arise.

Interdisciplinary work may yield positive benefits, of course, but Hvidtfeldt's point is that it is by means of the application of his approach-based framework that the various elements of such work may be disentangled and the benefits as well as the pitfalls then discerned.

There is, as he acknowledges, much more to be said, not least with regard to the application of this framework to other interdisciplinary case studies as well as to transdisciplinary work. Nevertheless, this book offers perhaps the first clear analysis of interdisciplinarity from the perspective of the philosophy of science and as a result, not only advances this particular debate but also sets out an entirely new direction for the field. In this respect, it represents a very worthy addition to the *New Directions* series.

Department of Philosophy, University of Leeds   Steven French
Leeds, UK

Preface

Before embarking on the somewhat complex discussions of this book, a few elucidatory remarks are in place by way of preface.

The book before you is about interdisciplinary science. It will be obvious to most that interdisciplinarity is quite a popular phenomenon in science today. One important question is "Why is interdisciplinarity this popular?" Another equally important question is "Is this popularity merited?" On the one hand, it appears that many activities, which fit the concept "interdisciplinarity" reasonably well, have delivered remarkable results. On the other hand, it is quite clear that we do not have good criteria for evaluating interdisciplinarity at our disposal. There is always a danger of being allured by phenomena which you do not know how to systematically assess.

With this book, I offer a framework for evaluating the extent to which particular cases of interdisciplinary science contribute to raising epistemic standards. The central contribution of this book is an application of recent and contemporary philosophy of scientific representation to cases of interdisciplinarity. This requires some adaptions to the philosophical framework as well as some discussion of how best to distinguish between different scientific approaches. The reward is a method, approach-based analysis, for assessing relevant epistemic aspects of cases of interdisciplinary science. The framework developed in this book is suitable for evaluating existing interdisciplinary projects. It may also, however, provide useful guidance for the ambitious practitioner of scientific crossbreeding.

The home of this project is in philosophy of science. Among people devoted to the study of interdisciplinarity, philosophy of science is not especially popular. Therefore, I believe, there is reason to assume that those engaged in what I refer to as 'Interdisciplinarity Studies' might not welcome my efforts with great enthusiasm. As I will discuss below, it is a broadly established "truth" in Interdisciplinarity Studies that philosophy of science is a redundant (and rather annoying) intellectual exercise, and that it has little relevance for actual scientific activities.

Without doubt, there are some core problems within philosophy of science. One is the trade-off between philosophical scholarship and genuine scientific expertise among theorists. In many cases, philosophers end up discussing issues which practitioners of the relevant types are actually in a much better position to handle (due to their deeper knowledge of the subject matter). On the other hand, many scientists are not motivated, and lack the training, for carrying out philosophical work with the conceptual rigour required. In some ways, then, philosophy of science is built on compromise.

It is my impression, though, that at least some ways of doing philosophy of science have unmistakeable utility. If I did not believe so, I would spend my time on something different. But philosophers should attempt to curb their propensity for trying to come up with a priori answers to questions which are essentially a posteriori in nature. One such question is, of course, whether or not philosophy of science contributes to developing science towards higher standards. This is, in the end, a matter for empirical research.

To handle such questions, we should, perhaps, establish a few new disciplines. A couple of suggestions could be "empirical meta-philosophy of science" and "science studies studies". We could categorise such enterprises as reflexive intradisciplinarity (intelligently?) designed to get to the bottom of what theorising about science is actually good for. It would be nice to have experts trained in these topics who could declare categorically ex cathedra that philosophy of science is immensely important.

Until that is established, I can only hope that reading (and writing) this book appears to be worth the effort.

This book has indeed required substantial effort and has been a long time in the making. Parts of the inspiration for this project popped up while I was a Masters student at the University of Copenhagen; other parts occurred to me while I was working as a research assistant at a psychiatric facility in the Capital Region of Denmark. Most of the central elements, however, were developed during my time as PhD student at the University of Southern Denmark. I would like to take this opportunity to thank a number of colleagues and friends for constructive discussions of some of the elements that make up this manuscript. These discussions have helped me considerably towards developing and refining the original raw ideas. While preparing this manuscript for publication, I enjoyed the encouragement and support of Department of Communication and Psychology at Aalborg University. Especially worth mentioning is my indebtedness to the Humanomics Research Centre generously funded by The Obel Family Foundation and The Velux Foundations.

To different extents and in different ways, the following have all provided helpful comments and suggestions as well as general encouragement along the way. They have each contributed in their way to the maturation and realisation of this project—probably without realising the full impact of their contributions. I am, obviously, grateful for all the inputs and inspiration.

I list in no particular order: Nikolaj Nottelmann, Signe Wolsgård Krøyer, Finn Collin, Jan Faye, Lasse Johansson, Mikkel Gerken, Simo Køppe, Esben Nedenskov Petersen, Jens Hebor, Mikael Vetner, Sara Green, Jonas Grønvad, Søren Harnow Klausen, Frederik Stjernfelt, Caroline Schaffalitzky de Muckadell, Jacob Luge Thomassen, Søren Engelsen, Cynthia M. Grund, Uffe Østergaard, Stig Børsen Hansen, Tom Børsen, Andreas Brøgger Jensen, Carl Bache, Lars Grassmé Binderup, Thomas Trøst Hansen, Peter Wolsing, Louise Amstrup, Jørgen Hass, Anne-Marie Søndergaard Christensen, Claus Emmeche, Emily Hartz, Joachim Wiewiura, Josef Parnas, Hans Siggaard Jensen, and David Budtz Pedersen.

During my stay in Sydney in the first half of 2015, I received helpful, inspiring, and (in some cases) even friendly comments and suggestions from, among others, John Matthewsson, Peter Godfrey-Smith, Dominic Murphy, Alan Chalmers, Paul Griffiths, Ofer Gal, Georg Repnikov, Sahar Tavakoli, Ian Lawson, and Chloe Collins.

I have further received helpful suggestions from a number of anonymous referees commenting on previous publications of mine.

Let me also express my gratitude towards my parents (without whom... and so on and so forth), all my brothers and sisters, brothers and sisters in law, and nephews and nieces.

Projects such as the one resulting in the manuscript before you take their toll—not just on the individual composing it, but also on the members of his or her nuclear family. In this respect, I must thank and apologise to my wife and our children: You have been incredibly patient and supportive even though your husband and father has been far more tense, preoccupied, and, indeed, physically absent than you (and I) would have preferred. I am fortunate and grateful!

Even though all the people mentioned (and forgotten) above have provided helpful comments and (direct or indirect) encouragement, I have no reason to believe (and considerable reason to doubt) that they or anybody else would approve this manuscript in its entirety. I take full responsibility for all the shortcomings exhibited in the following. It is clear that yours truly is solely to blame for anything that has made its way into this manuscript.

Copenhagen, Denmark   Rolf Hvidtfeldt

Contents

List of Figures

1

Introduction

'Interdisciplinarity'. What a wonderful buzzword.[1] On the one hand, there can be little doubt that drawing inspiration from various scientific approaches is central to developing and expanding our understanding of reality in the broadest sense. Indeed, the history of science is rich with cases of successful scientific achievements which can reasonably be considered interdisciplinary in one way or another. On the other hand, everybody has his or her favourite horror story featuring some obviously misguided (or even faux) interdisciplinary collaboration.

Curiously, however, very little effort has been put into the development of ways to distinguish between "good" and "bad" interdisciplinarity. In the words of Nancy Cartwright:

> Within each of the disciplines separately, both pure and applied, we find well developed, detailed methodologies both for judging claims to knowledge and for putting them to use. But we have no articulated methodologies for interdisciplinary work, not even anything so vague and general as the filtered-down versions of good scientific method we are taught at school. (Cartwright 1999, p. 18)

© The Author(s) 2018
R. Hvidtfeldt, *The Structure of Interdisciplinary Science*, New Directions in the Philosophy of Science, https://doi.org/10.1007/978-3-319-90872-4_1

There is no lack of academic interest in interdisciplinarity, though. Indeed, there is a large and growing literature on the topic. One might even speak of a virtual discipline of *Interdisciplinarity Studies*. But the Interdisciplinarity Studies literature is mainly focused on the social aspects of interdisciplinary collaborations, whereas the epistemic vices and virtues of interdisciplinarity are only rarely and cursorily discussed. Consequently, if we aim to understand whether and under which circumstances interdisciplinarity leads to beneficial epistemic results, we need to develop the required tools of assessment more or less from scratch. That is the primary goal of this book, then: To develop a framework, an articulated methodology, for evaluating epistemic aspects of interdisciplinary science.

To develop a more adequate way of capturing what is at stake in interdisciplinary science, I suggest drawing inspiration from the contemporary philosophical literature on scientific representation. A representation-based approach to the analysis of interdisciplinary science, and the discussion of the consequences of representing interdisciplinarity in this way, are the two main contributions I offer with this book.

To add one final, important qualification: This book offers a novel framework for analysing interdisciplinarity. Though it is a good one, it is only one of several possible and relevant frameworks. In the framework of this book, integrations of distinct scientific activities are idealised and represented in a certain way—one which emphasises aspects of interdisciplinarity which are out of focus in most other existing ways of analysing this phenomenon. That I focus on different aspects of the analysis of interdisciplinarity does not mean that I consider the standard approaches completely misguided. Still, I hope the reader will agree that viewing interdisciplinarity in the perspective developed below draws out interesting and relevant aspects, which may have the potential to alter the way in which we view interdisciplinary science.

The (Epistemic) Fundamentals of Interdisciplinarity

Let us start off with the following, somewhat banal observation: The concept "interdisciplinarity" presupposes, as a minimum, that some sort of *inter*-action and integration between at least two relevantly different parent *disciplines* takes place. Further, and at least as banal, there is a temporal aspect: "Interdisciplinarity" presupposes that there is a pre-interaction state of affairs in which the involved disciplines are distinct, and that there is a post-interaction, or integrated, state of affairs in which, unless the effort has been completely futile, some product of the integration of the parent disciplines has come into existence.

The basic idea in interdisciplinarity is, thus, to combine two or more scientific disciplines into some integrated approach (loosely speaking). The motivation for this kind of scientific crossbreeding is that through the combination of different scientific disciplines it might be possible to construct hybrids, which are in some respect superior to (at least one of) the parent disciplines.

Scientific quality is, of course, a difficult and controversial philosophical issue in itself and can be construed in many, quite different ways. Ultimately, determination of whether superiority has been achieved is, at least to some extent, dependent on the purposes the scientific enterprises in question are intended to serve. If one were to think of paradigmatic examples of improved scientific quality, reasonable examples might be increased explanatory power, the addition of detail or nuance, improved accuracy (for instance, in terms of prediction and/or distinction), improved reliability, improved validity, increased scope, more general implications, increased conceptual coordination, improvements in terms of cognitive economy (sometimes called 'simplicity'), or improvements in ability to intervene in relevant processes and produce, prevent, or control specific phenomena.

These are all[2] more or less standard textbook suggestions for evaluating scientific quality, which might all be relevant to discussions of epistemic enhancements due to interdisciplinarity. It bears emphasising once again that explicit discussions of how and to what extent interdisciplinary activities result in scientific or epistemic improvements are rarely encountered in existing treatments of the topic of interdisciplinarity.

In this book, then, the terms 'interdisciplinarity' and 'interdisciplinary science' are used to refer to scientific activities which involve integration of (elements of) theoretical representations picked from different scientific backgrounds. By epistemic assessment of interdisciplinarity, I mean the evaluation of how the integrated "knowledge" fares when evaluated along dimensions of scientific quality such as those listed above. This may reasonably involve a comparison with the epistemic vices and virtues of the parent disciplines.

Apart from epistemic issues, various other aspects of the activities involved in scientific practice may be considered good or bad by the involved scientists or other stakeholders. For instance, it is valuable to be able to maintain a living and it is quite attractive and very difficult to obtain (and retain) a job in academia. Consequently, one might expect that there is ample motivation for opportunistic interdisciplinarity. This possibility has received little attention in the existing literature—possibly because it presupposes a critical examination of whether interdisciplinary collaborations are implicitly good. Further, as the literature within Interdisciplinarity Studies clearly demonstrates, there are lots of non-epistemic issues relevant to analyses of interdisciplinarity. So, to be clear, even though focus is on epistemic aspects of interdisciplinarity in the following, other analytical dimensions should not be disregarded. Certainly, various kinds of aspects of scientific collaborations may have important implications for the development of science.

In this book, however, I primarily address epistemic aspects of interdisciplinarity. It is a central assumption of the book that epistemic aspects should be included in analyses and evaluations of interdisciplinary collaborations. It is a further assumption that interdisciplinary research activities, as other research activities, ought to be carried out cautiously and systematically in order to get the most out of the effort, while at all times maintaining a clear view for what benefits are gained through a specific effort. Throughout the book, I will provide examples which illustrate the predicaments one might end up in, if one's attitude towards certain epistemic pitfalls is too lax.

To be able to evaluate the extent to which particular cases of interdisciplinarity live up to the above-mentioned ideals, there are many issues which require considerably more attention than they usually get. If, as it is sometimes argued,[3] interdisciplinary work should be allowed to proceed in a less stringent manner than more traditional disciplinary science, at least there should be some sort of argument for why and in which respects such an attitude is considered beneficial. Indeed, such an argument might be carried out within the framework developed below.

Interdisciplinarity Studies

One of the basic reasons for developing an alternative approach to the analysis of interdisciplinarity is that epistemic issues are insufficiently dealt with in the existing literature on the topic. Despite all the merits of the Interdisciplinarity Studies literature, it does have significant short-comings, since a number of philosophical, and most pressingly epistemic, issues related to interdisciplinarity are largely unaddressed.

The absence of measures, or apparent attempts to develop measures, for the epistemic benefits of interdisciplinary collaborations may be partly due to that Interdisciplinarity Studies draw considerably on work by scholars from sociology and/or science studies. As science studies icons Collins and Evans have stated, "[t]he dominant and fruitful trend of science studies research in the last decades has been to replace epistemological questions with social questions" (Collins and Evans 2002, p. 236). There is no doubt that this trend has been dominant, and it has certainly also been successful—at least when measured in terms of popularity. But determining the extent to which it has been fruitful is, of course, a more difficult matter, which is closely related to the evaluation of science in general. I will argue that the focus on social aspects has blocked the light for (more) relevant epistemic concerns.

My Alternative

So, what is the alternative approach to the analysis of interdisciplinarity advocated in this book? As a starting point, I confront the conviction that conventional taxonomies of disciplines provide fruitful ground for analysing combinations of scientific approaches. I move on to suggest that a focus on *activities of representation*[4] reveals a much more interesting level of detail. A fundamental assumption of the argument below, an assumption which I take to be endorsed by a large group of influential contemporary philosophers of science (e.g. Cartwright 1999; Giere 2006; Godfrey-Smith 2009; van Fraassen 1980, 2008; Weisberg 2013), is that representation is an indispensable and central part of scientific activity. The further claim I make is:

> Representation is an indispensable and central part of scientific activity, and if interdisciplinarity has any significant effect on scientific practice, then the effect of interdisciplinarity must somehow be reflected in the representational activities as displayed in the products and outputs in post-interaction states of affairs.

"What are the products and outputs of science?" one might reasonably ask. For present purposes, my answer is this: Most tangibly, the products of science are the publications produced. But it is the propositional content of these publications that are of real interest. In my treatment, it is assumed that there are, basically, two relevant types of propositional content.

The first type of propositional content of scientific publications consists of (more or less) specified ways of representing (more or less) specified phenomena by means of (more or less) specified vehicles of representation. Sometimes a publication includes presentations of novel vehicles of representation; sometimes the central idea is an application of an established vehicle of representation to an object different from what has traditionally been targeted by means of the particular vehicle of representation applied. Finally, sometimes publications are about the re-application of a previously presented vehicle of representation (perhaps with certain adjustments) to a previously targeted object in order to reassess its value or previous results achieved (so-called replications). In the following, I use the expressions 'scientific approach' or simply 'approach' to refer to *a specific way of using a specific vehicle to represent a specific target*. How to explicate scientific approaches is going to be a central part of the machinery of this book. I will address this matter in more detail later in this chapter and return to different aspects of this complex issue throughout the book.

The second type of propositional content consists of the conclusions and recommendations which result from the analytical process in which vehicles of representation play a central part. I will refer to such conclusions and recommendations as *the outputs* of science, since those are often what make their way to the headlines of newspapers and thereby to the general public (and policy makers). It is noteworthy, though, that outputs are also what reaches most members of other parts of the scientific community. That is, though two parties both belong to the general scientific community, the one party rarely has deep insight into all the gory details of the activities of the other. This issue[5] will prove to be significant later on in this treatment.

So, the products of scientific activities are vehicles of representation and the specified ways of applying these as presented in publications. The other important kind of result of the activities of science, which could be considered a scientific commodity if you like, consist of predictions, recommendations, and interpretations supposed to constitute guides for action in the sciences as well as in broader society. Those *outputs* are derived from the representational activities and from analyses of the vehicles of representation involved. I suggest that it is fruitful to consider the outputs of scientific efforts as most often specifiable in terms of hypothetical conditionals or, for instance in cases where historical matters are analysed, in terms of counterfactual conditionals.[6] This issue will not be treated thoroughly in this book though, since the exact nature of scientific outputs is not central to the topics being investigated.

Does Everybody Represent?

Whether or not it is reasonable to choose representation as the focal point for the present analysis depends on whether representation is central, not just in some sciences, but in a relevantly similar sense in all scientific activities that might be involved in interdisciplinary activities. In this case, that means including scientific approaches traditionally categorised as belonging to the humanities and the health sciences as well as the natural and social sciences. Since attempts to introduce aspects of methodology from the natural sciences in, for example, the humanities are abundant, a level of abstraction is required at which the relevant aspects of all potentially involved disciplines can be incorporated.

My position is that such an understanding of scientific representation is attainable without straining generally accepted ways of conceptualising science beyond coherence. Indeed, many philosophers engaged in the debate on scientific representation would presumably agree, even though they rarely, if ever, discuss scientific representation in, say, the humanities.

In his seminal work on scientific representation, Bas van Fraassen states the following:

> Scientific representation is not exhausted by a study of the role of theory or theoretical models. To complete our understanding of scientific representation we must equally approach measurement, its instrumental character and its role. I will argue that measuring, just as well as theorizing, is representing. (van Fraassen 2008, p. 2)

For the present purposes, I stretch the concept of "representation" a bit further. As is common in philosophy of science, van Fraassen focuses on the most prestigious natural sciences.[7] However, the categorisations belonging to disciplines in, for instance, the humanities may (at an appropriate level of generalisation) be considered to be equivalent to the measurements of the quantitative sciences. The concepts of, for instance, literature theory are presumably less stringent and less well coordinated than the measurements of thermodynamics. But nevertheless, literature theorists use the concepts of literature theory to indicate that the conceptualised target has certain characteristics and plays a certain role in a larger theoretical scheme. Thereby, literary concepts fulfil the most basic requirement of van Fraassen:

> There is no representation except in the sense that some things are used, made, or taken, to represent some things as thus or so. (van Fraassen 2008, p. 23)

This is exactly what literature theorists do: They use some things to represent some other things (e.g. certain concepts used to represent characters in a novel [or vice versa]) as thus or so. Bas van Fraassen states, that if he were to propose a theory of scientific representation (which he stresses that he has no intention of doing), the above quote would be its *Hauptsatz*.

This "soft" attitude towards delineating scientific representation is in line with Mauricio Suarez, who states:

> I propose that we adopt from the start a deflationary or minimalist attitude and strategy towards the concept of scientific representation, in analogy to deflationary or minimalist conceptions of truth, or contextualist analyses of knowledge. Adopting this attitude [...] entails abandoning the aim of a substantive theory to seek universal necessary and sufficient conditions that are met in each and every concrete real instance of scientific representation. Representation is not the kind of notion that requires, or admits, such conditions. We can at best aim to describe its most general features. (Suarez 2004, p. 770 f.)

Van Fraassen and Suarez are right not to seek exact definitions of representation. And further, their quite inclusive accounts of representation[8] admits the treatment of a very broad class of scientific activities. On this background, I agree with van Fraassen that we can reasonably consider measurement to be representation, and I further add that so is categorisation. Once that is accepted, I believe we can reasonably answer affirmatively to the question of whether representation is central in a relevantly similar sense in all scientific activities that might be involved in interdisciplinary activities.

This would be in stark opposition to a widely-held position in which representation involves modelling, and 'model' is conceived as short for 'mathematical model' and therefore exclusively connected to the quantitative sciences. I agree on this issue with Thomson-Jones' (2012)[9] argument in favour of a propositional view of modelling. According to Thomson-Jones, most (if not all) mathematical models are somehow embedded in sets of propositions. These sets of propositions may for instance indicate how the mathematical structures of a model relate to its target system(s). On the other hand, examples of non-mathematical modelling consist solely of sets of propositions. The propositional view on modelling is especially useful in relation to an analysis of interdisciplinarity (such as the present one) in which one needs a way of conceptualising the vehicles by which "things are represented" that encompasses various divergent scientific approaches.

Thus, for present purposes I think it is reasonable to accept Thomson-Jones' claim that vehicles of representation are embedded in networks of propositions and that some instances of modelling do not involve mathematics at all. I will also claim, however, that for a meta-representation to be adequate, a finer level of detail is needed compared to what Thomson-Jones offers. Consequently, once the somewhat controversial move from "modelling" to "propositional modelling" is accepted (for the sake of argument at least), the next step is to attempt to spell out what these underlying propositional structures consist of.

One thing has to be kept clear, though. The claim that all scientific activities involve representation does not entail that all representational activities are scientific. To some, including concepts of the humanities in the class of scientific representation may seem to open a door for any layman concept to be included in this category as well. This, however, is a question for treatments of scientific demarcation, and poses no threat to our current investigation (which involves no such ambitions). If we are to be able to evaluate interdisciplinary science in all its forms, we had better not exclude anything for being less than truly scientific before we get started. Indeed, a considerable part of interdisciplinary science seems somewhat quasi-scientific at first glance.

The Intermediate Layer

On a naïve construal, vehicles might be believed to serve more or less as definitions which in themselves pick out which phenomena they are about. That is, the vehicle of representation could be construed as having the indexical function of pointing out its target build in somehow. This, however, cannot be the full story, since vehicles of representation are quite often transferred from one use to another. As one example, which I will return to below, Michael Weisberg has discussed how a mathematical model originally conceived to represent the dynamical relations between predators and prey in the Adriatic Ocean has (somewhat ironically) been used in economic theory to describe relations between different kinds of agents in the market (Weisberg 2013). It seems quite farfetched to claim that this second use was somehow already pointed out by the model in its original formulation. Rather, there must be something else mediating the relation between vehicle of representation and target in any representational activity.

On the other hand, a specific vehicle of representation certainly picks out a set of possible states. In other words, a vehicle puts constraints on which results, given certain inputs, we shall expect from the targeted phenomena. Otherwise, of course, the representation would not tell us much. Vehicles do, in this way, perform a very central conceptually limiting task. The set of possible states which the vehicle of representation can display frames our understanding of the target system.

Tools, Algorithms, and Basic Assumptions

The vehicle of representation is relatively easy to identify, but, unfortunately, it is only the tip of the iceberg when considering representation. The vehicle of representation is part of, admittedly a very central part of, what I refer to as the 'approach', which, as already stated, is going to be one of the most central terms in the discussions of this book. But before we can get a hold on how I construe approaches, we need to discuss the topic of what I call *the intermediate layer* between vehicle and target.

Following van Fraassen, there is no representation unless something is used[10] to represent something else. But what does 'use' mean? To analyse representational activities, we need to get a hold of what use is in the present context. The question of "how a vehicle is used" can, I believe, fruitfully be replaced by the question of "how the vehicle is linked to its target". In the analyses of this book, I construe the connection between vehicle and target as constituted by an intermediate layer consisting of combinations of more or less explicit, more or less taken for granted, assumptions and (conceptual) tools of various kinds.

My take is the following: Supporting the representational activities is first of all a group of fundamental assumptions (which I take to constitute at least a significant part of the propositions which Thomson-Jones discusses). Second, the representational activities are supported by a group of tools, which do not represent anything themselves, but which serve various other purposes which are central for establishing the connection between vehicle and target. One such function is to translate raw inputs into data (in terms of concepts or figures) which can be processed further by means of other tools, until the link to the vehicle of representation is established. Some of the tools involved are literal tools (for instance various kinds of more or less complex instruments), others are mathematical tools like statistical methods. Further, I suggest that another subgroup of the tools involved can be fruitfully construed in terms of *propositional algorithms*, by which I mean (more or less explicitly stated) sets of rules for carrying out certain conceptual operations (an example will follow soon below). Involved in linking vehicles to targets is also a number of sub-representations, including what is often referred to as 'data models'. Data models are ordered groups of data represented in ways appropriate for a certain purpose. Data models can be analysed in order to derive inputs to feed other elements in the representational chain. The results of processes of measurement and categorisation also count as sub-representations.

While vehicles and other sub-representations are used for representation, algorithms, tools, and assumptions which contribute to linking vehicle and target may not represent anything. They are means for linking target and vehicle, but they are not necessarily intended to represent any *real* connections between the two.

The claim that some of the tools in the intermediate layer could be characterised as propositional algorithms needs further underpinning. As an example, in any representational activity one needs some way of pointing out the phenomena in focus. One way of doing this, though by no means the only way, is to use one of a number of possible types of definition. A type of definition is, I believe, a good example of a propositional algorithm doing important supporting work in representational activities. Different types of definitions have different conceptual structures, which again can be characterised as differently structured sets of rules for deciding whether something falls under a concept or not. In my use of the term, each specific set of rules for carrying out a conceptual operation would be a specific propositional algorithm. I will spell out and discuss the algorithms of different types of definition in Chap. 6.

As an example of a propositional algorithm for picking out objects of interest without using definitions, one might consider a setting in which the cognitive system of a conceptually well-functioning individual is sufficiently accurate to point out phenomena of interest. An example of a psychological study of the effect on elderly people's well-being from owning a dog as compared to owning a cat or a canary will be discussed in the section below. In such a setting an exact definition of "dog" would not be required. Instead the propositional algorithm might be something like: *let a person with a normal understanding of the words 'dog', 'cat', and 'canary' determine whether the elderly person in question owns one or the other (or perhaps no pet at all).*

Both types of algorithms (definitional and non-definitional) discussed above generate sub-representations in terms of concepts which can be further processed by other tools.

Other candidates for the status of propositional algorithm might be: Different ways of idealising, different ways of abstracting, different ways of measuring, different ways of observing, different ways of categorising, different ways of coordinating basic concepts, different ways of gathering data, different ways of quantifying data, different ways of using statistics (including the choice between specific statistical approaches), different ways of analysing topics or data, different ways of creating graphs and diagrams, different ways of interpreting graphs and diagrams, different ways of setting up experiments (think of standards such as randomised double-blinded studies), different ways of intervening or not, or different ways of creating taxonomies.

An Example

Since the above few paragraphs may come across as quite abstract, let me offer the following constructed scenario by way of example.

Let us say that a social psychologist wants to study effects of owning different kinds of pets on the wellbeing of single, elderly persons. Let us say that the psychologist operates with the equation in Fig. 1.1 as a vehicle for representing the assumed "pet-effect".

In Fig. 1.1, x is the weight of the pet measured in kilograms and w is the weight of the owner. Further, g is a measure for the grumpiness of the elderly person, whereas c is an estimate of how cute the given type of pet is (g and c are measured on each their scale of 1 to 10).

$$pet\text{-}effect = \frac{\left(1 + \frac{g}{\pi}\right)^{-\frac{1}{2}\left(\frac{x+\pi-\frac{w}{x}}{2}\right)^2}}{\frac{1}{c}\sqrt{2\pi}}$$

Fig. 1.1 The pet-effect

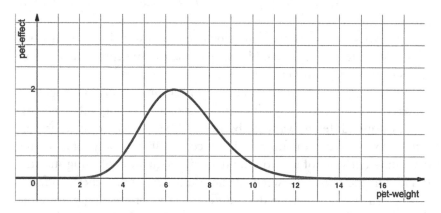

Fig. 1.2 The pet-effect graph

One might interpret this equation quite easily by means of a graph. If one assumes that a pet-owner weighs 60 kilograms and is quite grumpy (say, with a score of 8), whereas the pet is about averagely cute (5), one will reach the optimum pet-effect with a pet weighing just about 6 kilograms (as can be read of Fig. 1.2 below). The cuter the pet, the stronger the effect. The heavier the owner, the higher the optimal weight of the pet. The grumpier the owner, the narrower the range of pet-weights with a positive effect.

Now, something is needed to mediate the connection between this equation (and graph) and the reality of pet-ownership and wellbeing. Among these would be:

Various tools (such as):

- Stipulative definition of what 'elderly' means in the context as a means to pointing out a relevant sample.

- Normal functioning human category system to determine whether people in the sample own a pet, and which kind of pet it might be.
- A scale to measure the weight of the pets and people.
- A set of operationally-defined categories to measure the grumpiness of the elderly person.
- Multiple-choice questionnaires to measure the individual elderly persons' experiences of wellbeing.
- Mathematical tools helpful for quantifying the raw (qualitative) data.

Various assumptions (such as):

- Assumptions about the cuteness of different kinds of pet.[11]
- Technical assumptions such as that $\alpha \leq 0.05$ is a reasonable threshold for statistical significance.
- That wellbeing is something you can measure and quantify in a meaningful way.

By means of these and other elements, it is possible to link the vehicle (the above equation) with the phenomena of interest. This activity, when interpreted, might result in an output along lines such as:

> By analysing our model of the pet-effect, we conclude that living with a medium sized reasonably active pet that requires some physical interaction has a positive effect on your well-being by forcing you to do low-intensity exercise. The conclusion is (in the form of a hypothetical conditional): "If you are an elderly person and you want to improve your well-being, you should get the equivalent of a cute, medium sized dog".

Should our friend the psychologist decide to get involved in interdisciplinarity for some reason, opportunities are plenty. He might be inspired by studies in biology to control what is going on in a stricter way. He might add assumptions that dogs in these kinds of studies should be bred to have similar levels of activity and temper. He might decide to isolate his specimens in a laboratory. He might decide to collaborate with neurologists to develop a deeper understanding of grumpiness by fMRI-scanning his subjects. Or he might even conspire to use specially bred elderly people to study the effect of grumpiness in a more controlled setting. He might import statistical tools or other ingenious ways of analysing quantitative and qualitative data. Or he might import some existing equation, the mathematical structure of which fits his target (or intuitions) even better.

All such interdisciplinary activities should result in changes in how the phenomena of interest are represented. One should be able to discern these changes by looking at details such as those sketched out above, that is, at changes in vehicle of representation or among the elements constituting the intermediate layer.

Why Engage in this Kind of Madness?

My attempts to spell out details about which tools and assumptions are used in a given approach are motivated by the following assumption.

The analysis of interdisciplinarity in terms of combinations of representational approaches involves in its most basic form the transferral of vehicles of representation from some setting to some other setting. This means importing existing vehicles of representation and applying them to a different target or, perhaps, the hybridisation of vehicles already in use and (parts of) imported vehicles to create new ones. Thinking of interdisciplinarity in terms of combinations of models or vehicles of representation is far from adequate, though. The vehicles of representation are the proudly presented figureheads by which the phenomena of interest are represented. But, as already stated, vehicles of representation do not inherently represent anything in themselves. Therefore, they need to be embedded in larger networks of supporting elements to perform their representational magic. All aspects of the network supporting the representation are candidates for being transferred between scientific approaches as well as the vehicles themselves. Unfortunately, these networks are not necessarily explicitly stated in publications since a lot of the involved assumptions and tools are parts of the (tacit) background knowledge of people specialised in a given field of study.

Disciplines, thus, might be integrated in much more subtle ways than the transferral or combination of vehicles. If we are to do interdisciplinarity justice, we need to dig deeper into how representation is accomplished in the pre- and post-integrated states of affairs. And, indeed, it is required to spell out the difficulties involved when combining elements from different disciplines.

The last remark in the previous paragraph refers to one significant complication not yet mentioned. In the philosophy of scientific representation, it is frequently argued that representation is not neutral and that different perspectives on the same target may be incompatible. Put in another way, representation involves distortion. And whether or not the inherent distortions are tolerable is dependent on context. No tools, assumptions, or vehicles come certified for general use. In the present context, we therefore need to discuss the difficulties with combining non-neutral, incompatible perspectives. There are good reasons to assume that individually distorted elements picked from two distorted perspectives do not necessarily add up to something less distorted.

Adopting the framework sketched out above will lead us to a quite non-standard conception of interdisciplinarity. In the rest of this book, I will provide much further detail as well as concrete examples. Hopefully, this will eventually convince the reader of the usefulness of approach-based analysis of interdisciplarity.

A Bit of Terminological Explication

Before we venture into the main part of this book, it will be beneficial to settle some terminological issues. For the following discussion, we need terms for referring to classes of entities, some of which are not commonly distinguished in the literature and therefore have no commonly accepted names. A number of these have been introduced above, but it will be beneficial to have a more detailed discussion of the most important ones.

Perhaps the easiest route to grasping the first of these categories goes through an analogy with Hempel's and Oppenheim's distinction between 'explanans' and 'explanandum'. In the context of a discussion of scientific representation, one might be prompted to talk of repraesentans (= "*that which is used to represent*") and repraesentandum (= "*that which is represented*").[12] In the existing literature on scientific representation, the expression 'target system', or simply 'target', seems well suited for picking out what is being represented. This is good fortune since it would be nice to avoid such heavy terms as 'repraesentans' and 'repraesentandum'. Importantly, a target system is not (always) simply a part of the world but often a specific idealised or abstracted construct of a given phenomenon, set of phenomena, or, for that matter, fictive objects. There are no principled limitations as to what it is possible to represent.[13] Nevertheless, the conclusion is that 'target' or 'target system' fits neatly into the analysis below.

On the other hand, there is no neat term in the literature for the class that would be picked out by "repraesentans". As will be discussed in more detail below, models and modelling have attracted most of the attention since Ronald Giere set the stage for the contemporary philosophical discussions of scientific representation in (1988). However, as recent discussions have pointed out, far from all scientific representations are realised by means of models (Weisberg 2007, 2013). Modelling is understood as the practice where analyses, interventions or experiments are carried out, so to speak, by proxy. That is, in modelling the analytical efforts deliberately involve the use of an explicitly constructed mock-up of some kind representing the target. Activities referred to as *abstract direct representation*, in contrast, are characterised by using theoretical constructs to represent some target or target system directly without involving modelling activities. The work of Darwin and Mendeleev are oft-mentioned canonical examples of science allegedly proceeding by abstract direct representation, in that they refer directly to real world phenomena and their characteristics. Further, many other kinds of things (such as pictures or verbal descriptions) are used for representational purposes in science, apparently without being instances of modelling (at least not in any remotely obvious sense).

Due to this heterogeneity, it is useful to have a generic term for "all things which are used to represent a target system". In this book, '*vehicle of representation*' (or simply 'vehicle' for convenience) is used to refer to the "thing" that takes up the central place as *that which is being used to represent something else* in a scientific context.

As just stated, in the philosophical literature on representation, models are by far the most studied type of vehicle by which other things are represented (Cartwright 1983; Giere 1988, 1999, 2006; Godfrey-Smith 2009; van Fraassen 1980, 2008; Weisberg 2013). However, being the central term in an extensive literature also means that the exact sense of the word is quite controversial (Godfrey-Smith 2006, p. 725). And stipulating yet another (in this case very broad) sense of the term is bound to cause controversy (if not outright anger) as well as confusion. I will use the word 'model' frequently throughout the book. But this use will be restricted to cases in which modelling (in a narrow sense) is an obvious part of the representational activities. Most of the time the generic expression 'vehicle of representation' (or 'vehicle') is more fitting.

So, a vehicle of representation may be a mathematical model, a computational model, a propositional model, a concrete model, a theory, a linguistic expression, a concept, a painting, a piece of music, an open cheese sandwich,[14] or whatever. Contrary to what one might initially think, vehicles of representation used in a given act of scientific representation are much easier to pinpoint than target systems. A vehicle is most often given a prominent place in the publications of those making use of it. Indeed, one way to determine what is the vehicle of representation in a given approach is to look for what is proudly presented as the condensed "essence" resulting from the efforts on which a given scientific publication is based. In contrast, it often requires significant effort to explicate the target system.

To borrow an example (Fig. 1.3) from Ronald Giere (1988, p. 70 ff.):

$$p = 2\pi \sqrt{l/g}$$

Fig. 1.3 The simple pendulum

is an equation (a mathematical model) commonly used in physics textbooks to represent a certain aspect of the movements of a pendulum. The equation fits the experimental results of Newton and Galileo, which showed that the period (P) of a pendulum is proportional to the square root of its length (l) divided by the gravitational constant (g), and that the period is independent of the mass of the bob (which is, as a consequence, not represented).

In this case, it could hardly be more simple: The above equation is the vehicle of representation in the mentioned act of representation.

Targets

The target, however, is less straightforward to identify. While vehicles often come in the form of mathematical equations, graphs, or verbal descriptions easily put to print, target systems are much more diverse. A target system might be a physical object, a group of physical objects, or a (stipulated) kind of physical object. It might also be an emergent object, a fictional object, a social object or relation, a process, or, for that matter, a more or less universal scientific practice. But to complicate things further, in acts of representation it is quite rare simply to represent some object per se. Mostly, some aspects of the target is represented while others are (more or less deliberately and more or less explicitly) ignored.

The target in the case of the equation of the simple pendulum is *not* any old pendulum in motion. In this case, we are lucky since it is most often made clear in physics textbooks that the above equation represents an idealised pendulum. Indeed, it is often made explicit which idealisations are required in order to single out the kind of target the equation fits. In the case of the equation of the simple pendulum, the target is *explicitly abstract*, indeed, explicit to a highly commendable degree. In other cases, of course, the target may either be (closer to) a real-world system, or more or less abstract without the involved idealisations being explicitly stated.

Approaches

As mentioned above, a central part of my treatment in the following will be an attack on the use of distinctions based on disciplines. I argue, perhaps at first glance paradoxically, that in order to understand interdisciplinarity, we will have to discard the notion of "discipline" and study approaches instead. Disciplines are highly complex phenomena and an entire chapter of this book is devoted to capturing their nature. Indeed, as the discussion of disciplines will reveal, disciplines are far too complex phenomena to serve as the basis for the epistemic assessment of interdisciplinary cases.

Therefore, I will focus on my alternative concept: "approaches".

As already stated, I define "approach" *as the specific way of using a specific vehicle to represent some target.* In a given publication, it is relatively straightforward to see *who* is representing, *what* is used as a vehicle of representation, *what* is represented (with the qualifications discussed just above), and *why*. Further it is easy to get a decent idea about what is supposed to be the consequences of the approach, that is, what the outputs are.

To return to the stylised example presented above, in a psychology paper the number of participants, the statistical methods used, the questions asked, and so on will (at least ideally) be spelled out. If experiments are carried out, the specific experimental setup will be described. Some aspects will not be explicitly presented, though. Some assumptions and use of tools will be taken for granted. Lots of details will not be included, such as biographical details of the participants in the experiment. Some of these omitted details might matter if closer analyses are carried out, of course. The approach includes everything that is presupposed by those doing the representation, including implicit assumptions and other very general and often unquestioned principles (such as parsimony assumptions and similar).

Some approaches are very complex. But compared to disciplines, they are much easier to handle analytically. Inspired by Giere's suggestion that we should consider theories to be clusters (of clusters of clusters) of models

(Giere 1988, p. 82), I suggest that we should consider disciplines as bundles (of bundles) of approaches.

I intend my use of 'approach' to be in some ways similar to everyday (scientific) parlance, in which it would not be surprising to encounter questions such as "What is your preferred approach?", "What was their approach for tackling that issue?", or perhaps "This doesn't seem to work—let's try a different approach!" Indeed, what I wish to capture are the changes that constitute transformation from one approach to another, since this is how I believe interdisciplinarity should eventually be considered: as transformation from one approach to another. Every approach requires the same kinds of appraisal in order for its epistemic virtues to be assessed.

"Distance" and "Proximity"

Independent of which people are involved in the scientific activities and their respective affiliations, different approaches may be more or less overlapping in terms of sharing vehicles and other elements of intermediate layers. On an approach-based account, two approaches may well be considered very proximate due to strong overlaps in the above sense as well as for sharing paradigmatic examples of good practice in spite of belonging to what is traditionally considered to be very different disciplines. Though far from being accurately quantifiable, "*Distance*" and "*proximity*"[15] are two concepts (denoting opposite limits of a continuum) meant to refer to the degree of overlap between two approaches.

Integration between proximate disciplines are similar to what is sometimes referred to as *narrow interdisciplinarity* in the Interdisciplinarity Studies literature, while integration between distant disciplines are referred to as *broad* or *wide interdisciplinarity* (Klein 1990, p. 18). I believe, that there is ample room for further elaboration of the notions of "proximity" and "distance". Especially, my analysis will reveal interesting pitfalls of trying to integrate distant approaches.

On the one hand talk of interdisciplinarity between proximate approaches could in many ways seem quite unimpressive from a naïve epistemic point of view. The combination of two, or more, very similar approaches may intuitively seem unlikely to lead to revolutionary new insights. On the other hand, there might be overwhelming difficulties related to combining very distant approaches, since such would have very little in common at the outset. Thus, the required effort for solving the puzzle of whether such a combination could lead or has led to fruitful results might not even be worth the effort.[16]

Summing Up

To sum up the general theme of the discussions of this book: If interdisciplinarity has any epistemic effect worth noticing, at least a significant part of this effect must occur as a detectable effect at the level of scientific approach, that is, at the level of vehicles of representation or ways of using a vehicle of representation to represent a certain target for representational purposes. The analysis of such effects will include explication of the basic assumptions, tools, propositional algorithms, and data models at work in the pre- and post-integrated representational activities.

In the following analyses, interdisciplinarity will be understood as minimally requiring the integration of two approaches. Integration is understood as the combination of elements picked from different approaches. Distance between approaches is understood in the soft sense of a measure of the extent to which two approaches share elements— some of which might be more important than others, of course. In many cases of interdisciplinarity, one might expect that elements are picked from several (more or less closely related) approaches.

Should one encounter an alleged interdisciplinary collaboration in which no differences can be detected at the levels of approach, it ought to make one somewhat suspicious of faux interdisciplinarity. If, on the other hand, nothing seems to have changed in terms of vehicles, targets, and relations between them, while differences in output still occur, it would indicate that my presented framework fails to capture the most central elements of interdisciplinary integration.

What has been presented in this introductory chapter so far is the basis upon which I will develop a framework for analysing the epistemic aspects of interdisciplinarity—the so-called *approach-based analysis*. The further development hereof will proceed in the following steps:

Chapter 2—Disciplines and Approaches

In this chapter I provide an in-depth analysis of the concept of "scientific discipline" and disciplinary difference.

As already stated, a basic point of attack in this book is conventional disciplinary distinctions. This chapter disentangles a number of dimensions of disciplinarity and thereby provides reasons to assume that a conventional scientific taxonomy is not a good basis for analysing epistemic aspects of interdisciplinary science.

On this background, it is argued that the concept of "approaches" is a more fruitful alternative.

Chapter 3—Interdisciplinarity Studies and Interdisciplinarity

I will provide a number of examples of interdisciplinary science and go through some relevant aspects of the present state of the discipline of Interdisciplinarity Studies.

I will provide my best account of why interdisciplinarity is such a topical subject in academia today, and why I consider the existing literature on the topic to be lacking in certain central respects.

I will also discuss the basic question of why we need interdisciplinarity at all. What are the attractions (the ontological and methodological assumptions) that drive people towards interdisciplinary collaborations?

I will further discuss different modes of disciplinary integration, from the polymath-mode (where a single person integrates several approaches internally) to entirely social modes (where the integration is, so to speak, external to the individuals involved). This discussion draws on discussions from science studies to explicate potential difficulties such as the lack of core sets on the one hand as well as the lack of peers to review publications on the other. Even though one might criticise the discipline-based peer review system for many reasons, it certainly does have a large number of points in its favour. Some of these positive aspects of peer review might be ineffective when addressing interdisciplinary approaches, though.

Chapter 4—Is Philosophy Relevant to the Study of Interdisciplinarity?

This chapter will be quite short. My answer to the above question will be "Yes!"

It is worth discussing, however, the quite prevalent and influential arguments against the relevance of philosophy to scientific activities (in general).

Three different perspectives for philosophical analysis of interdisciplinarity are discussed: (1) Patricia Kitcher's scholarly, retrospective approach, (2) Michael Weisberg's model-based approach (which, strictly, is not an attempt at analysing interdisciplinarity but nevertheless ends up discussing some of the central issues), and (3) my own more detailed approach-based analysis.

Chapter 5—Representation

In the remainder of the book I will argue in favour of construing interdisciplinarity and interdisciplinary activities by means of concepts and theories drawn from the philosophy of scientific representation. To develop this argument, it is required to introduce this literature in some detail.

I will go through discussions of the nature of models, not because I believe "model" is the perfect concept to place at centre stage in this discussion, but because the discussion of models throws valuable light on how best to construe interdisciplinarity qua representational activity.

Ronald Giere's framework for analysing representational activities is presented. The virtues of this approach are highlighted as well as the senses in which it is lacking in order to serve the present purpose.

Chapter 6—Pluralisms, Perspectives, and Potential Problems

In this chapter, issues related to scientific pluralism are discussed. This includes Giere's perspectivism and his one-world-(working-)hypothesis.

Scientific pluralism denotes the conviction that there are numerous ways to perform scientific activities and, importantly, these are not readily interchangeable between contexts.

Importantly, any way of representing reality involves distortions. And these distortions cannot easily be removed, if at all. Several in-depth examples of unavoidable distortions are presented. In many cases, such distortions are invisible to the untrained eye, which is a serious epistemic problem in interdisciplinarity, since interdisciplinary activities by definition involve lots of less than well-trained eyes.

At this point the concepts of "tools", "assumptions", and "propositional algorithms" will prove their worth. I go into the nitty gritty of some examples of importing tools and propositional algorithms and the pitfalls involved in such operations.

This chapter is long and deals with quite complex issues. I nevertheless recommend reading the entire thing, since it contains important discussions of central aspects of our present topic. Readers with little or no philosophical training may find it somewhat laborious to get through, but it will be worth the effort.

Chapter 7—Scientific Crossbreeding

In this chapter *the Giere Duplex* is introduced in detail as a framework for analysing interdisciplinary integration.

With the Giere duplex in place, we are in a good position to analyse integrations of scientific approaches. Such analyses may contribute to the systematisation of the considerations about modes and difficulties of integration discussed in Chap. 3.

Further, analyses based on the Giere Duplex lead to a richer understanding of the notions of "approach" and "distance". They also lead to some recommendations that run directly counter to established assumptions about interdisciplinarity.

Chapter 8—Case Study: Phenomenology Imported with EASE

In this chapter, the developed framework is applied in a case study of a specific interdisciplinary project within schizophrenia research, namely a particular approach of the EASE project developed by a research group lead by Josef Parnas at a Psychiatric Research Facility in Copenhagen.

I go through the background of the project, which includes the contemporary status of schizophrenia research, psychiatry, and psychopathology.

Parent approaches are singled out (to the extent possible), the integrated vehicle is identified as well as the most important elements of the intermediate layer. The approach-based analysis of this interdisciplinary project will result in interesting conclusions about the epistemic merits of this particular interdisciplinary approach as well as interesting, more general reflexions.

Chapter 9—Conclusion

In Chap. 9, I sum up the entire discussion and consider the consequences of viewing interdisciplinarity in the suggested way. I briefly discuss some possible ways to develop and further support the suggested framework as well as questions which would be good to address in future research.

Notes

1. Throughout this manuscript (except for this note) 'single inverted commas' are used when referring to words, whereas "double inverted commas" are used when referring to concepts. Following one of several Scandinavian traditions "double angle quotation marks" are used for in text quotations as well as scare quotes.
2. Depending on one's favourite textbook.
3. An anonymous referee commenting on an early version of (Hvidtfeldt 2017) responded to my claim that interdisciplinary activities ought to be carried out "cautiously, systematically, and stringently" with the comment: "hm… but you know, in ID it is precisely the opposite attitude that you need…".
4. Chapters 5, 6, and 7 are devoted to a thorough discussion of the philosophy of representation and the adaptions I consider to be required for a representation-based account of interdisciplinary science.
5. Which has, indeed, been addressed empirically within science studies (e.g. Collins 1981, 1985).

6. The idea of construing outputs of science in terms of conditionals is inspired by ideas presented by Peter Godfrey-Smith in a series of lectures at The University of Sydney in 2015.

7. Arguably, physics, chemistry, and biology are the places to make your mark if you want to "be someone" in contemporary analytical philosophy of science.

8. Actually, quite a bit more inclusive than either of them considers explicitly, I believe.

9. I will discuss this in some detail in Chap. 5.

10. The attentive reader will remember that van Fraassen's requirement is that something is "used, made, or taken" to represent something else. I will focus on use in the following, since I take it that "use" may function as a generic concept under which "made" and "taken" can be subsumed.

11. A dog you will have to take for a walk (= significant activity demand). A cat you will have to poke with your walking stick when it is about to urinate on the carpet again (= medium activity demand). A canary you will only have to feed occasionally, and flush in the toilet when it dies (= low activity demand).

12. As is well known, the original formulation runs as follows: "By the explanandum, we understand the sentence describing the phenomenon to be explained (not that phenomenon itself); by the explanans, the class of those sentences which are adduced to account for the phenomenon" (Hempel and Oppenheim 1948, p. 136 f.).

13. Once we get to my discussion of the elaborated duplex version of Giere's representational relation in Chap. 7, I will assume that the W in Giere's formalisation refers to target systems rather than real world aspects.

14. This will, I promise, eventually make sense.

15. I thank John Matthewsson and Peter Godfrey-Smith for independently pointing out the need for some notion of "disciplinary distance" in discussions of this project during my stay at The University of Sydney in 2015.

16. I disagree, then, with a lot of the existing treatments of interdisciplinarity in which it is assumed that disciplines have a "virtual monopoly of expertise on their subject matter" (Weingart 2010, p. 9). In a lot of cases such a generalisation would not pass an empirical test. This superficial view, I claim, is an artefact of uncritically basing ones understanding of interdisciplinary integration on conventional disciplinary taxonomies.

References

Cartwright, Nancy. 1983. *How the Laws of Physics Lie*. Oxford; New York: Clarendon Press; Oxford University Press.

———. 1999. *The Dappled World: A Study of the Boundaries of Science*. Cambridge, UK; New York: Cambridge University Press.

Collins, Harry M. 1981. The Place of the Core-Set in Modern Science: Social Contingency with Methodological Propriety in Science. *History of Science* 19 (1): 6–19.

———. 1985. *Changing Order: Replication and Induction in Scientific Practice*. London: The University of Chicago Press.

Collins, Harry M., and Robert Evans. 2002. The Third Wave of Science Studies: Studies of Expertise and Experience. *Social Studies of Science* 32 (2): 235–296.

Giere, Ronald N. 1988. *Explaining Science: A Cognitive Approach, Science and Its Conceptual Foundations*. Chicago: University of Chicago Press.

———. 1999. *Science without Laws, Science and Its Conceptual Foundations*. Chicago: University of Chicago Press.

———. 2006. *Scientific Perspectivism*. Chicago: University of Chicago Press.

Godfrey-Smith, Peter. 2006. The Strategy of Model-Based Science. *Biology and Philosophy* 21: 725–740.

———. 2009. Models and Fictions in Science. *Philosophical Studies* 143: 101–116.

Hempel, Carl G., and Paul Oppenheim. 1948. Studies in the Logic of Explanation. *Philosophy of Science* 15 (2): 135–175.

Hvidtfeldt, Rolf. 2017. Interdisciplinarity as Hybrid Modelling. *Journal for General Philosophy of Science* 48 (1): 33–57.

Klein, Julie Thompson. 1990. *Interdisciplinarity: History, Theory, and Practice*. Detroit: Wayne State University.

Suarez, Mauricio. 2004. An Inferential Conception of Scientific Representation. *Philosophy of Science* 71 (5): 767–779. https://doi.org/10.1086/421415.

Thomson-Jones, Martin. 2012. Modeling without Mathematics. *Philosophy of Science* 79: 761–772.

van Fraassen, Bas C. 1980. *The Scientific Image, Clarendon Library of Logic and Philosophy*. Oxford; New York: Clarendon Press; Oxford University Press.

———. 2008. *Scientific Representation: Paradoxes of Perspective*. Oxford; New York: Clarendon Press; Oxford University Press.

Weingart, Peter. 2010. A Short History of Knowledge Formations. In *The Oxford Handbook of Interdisciplinarity*, ed. Robert Frodeman, Julie Thompson Klein, and Carl Mitcham. Oxford: Oxford University Press.

Weisberg, Michael. 2007. Who is a Modeler? *British Journal for the Philosophy of Science* 58: 207–233.

———. 2013. *Simulation and Similarity: Using Models to Understand the World. Oxford Studies in Philosophy of Science.* New York: Oxford University Press.

2

Disciplines and Approaches

In this chapter, I provide an analysis of what the "things" called disciplines are. The core of the chapter will be discussions of various concepts of disciplinarity. There is a lot to say about the "nature" of disciplines, and most of it is interesting and highly relevant for discussions of interdisciplinarity. The perspective on these issues I present in this chapter will frame and inform the analyses in the subsequent chapters. Most importantly, the discussion of disciplinarity in this chapter will support my recommendation of analysing cases of interdisciplinarity in terms of approaches.

Disciplines are far more complex phenomena than the ordinary use of the word 'discipline' admits. One consequence of this is that if one wants to consider epistemic aspects of interdisciplinarity, one cannot simply rely on 'discipline' in the common usage of the word. In many cases, lexical definitions do not capture the complexity of the phenomena in target. Such discordances frequently cause widespread confusion and disagreement. This is certainly the case with the word 'discipline'.

The topic of scientific disciplinarity is sufficiently rich and interesting to provide material for a book-length treatment in itself. In this context, the treatment of these questions will have to be restricted to the most relevant aspects. Consequently, this chapter will not provide detailed

© The Author(s) 2018
R. Hvidtfeldt, *The Structure of Interdisciplinary Science*, New Directions in the Philosophy of Science, https://doi.org/10.1007/978-3-319-90872-4_2

accounts of the history of discipline formation or organisational analyses of how academic institutions have developed into their present form. Thus, the discussions in this chapter will by no means constitute an exhaustive account of disciplinarity.

The following discussion will quickly reach the conclusion that the concept "discipline", in any of the shapes analysed, is a poor foundation for analysing epistemic aspects of poly-disciplinary activities. At first, then, it might seem somewhat paradoxical that I will devote an entire chapter to this analysis, and further continue to use the terms 'discipline' and 'interdisciplinarity' throughout the book. Not least since I will argue that phenomena commonly referred to as interdisciplinary are more fruitfully discussed if analyses are based on the integration of approaches in my sense of the term.

But since 'interdisciplinarity' is the term by which the phenomena of interest are usually denoted, it will be too awkward to discuss changes to our understanding of interdisciplinarity without using the commonly accepted term. And further, though I reject "discipline" as the basis for analysing specific cases of interdisciplinarity, I do not claim that the concept is generally useless or superfluous. It certainly makes good sense to speak of disciplines in many contexts. And, indeed, a nuanced understanding of disciplines is required, in order to understand many of the issues related to approach-based analysis of interdisciplinarity. Acknowledging the terminological messiness, however, I will take care consistently to use 'approach', 'discipline', and 'interdisciplinarity' in ways that should reduce confusion to a minimum.

In order to carry out a research project at all, you need a scientific base. Most often, interdisciplinarity is practised from within some discipline and certain fundamental requirements of this discipline will have to be accepted. As will be discussed in Chap. 8, if you are aiming to publish your findings in psychiatric journals you cannot simply discard official diagnostic procedures. As that discussion will show there is a large difference between importing elements of philosophy into psychiatry and vice versa.

In most cases, a lot of background assumptions will not be explicitly challenged during integration. And if unchallenged, they are likely to remain in the default position of the base discipline. All in all, the concept of "discipline" is far from superfluous even when approaches are considered the basic unit of integration. Indeed, it will often be quite informative to know the disciplinary relationships of an analysed approach.

The main purpose of this book is to develop a method for singling out the epistemic effects and dynamics of interdisciplinarity. To do so, a fine-grained procedure for distinguishing between distinct scientific activities is required—among other things, in order to be able to determine what is combined in cases of interdisciplinary integration. For this purpose, I argue that my concept of "approach" is superior to "discipline".

As the following discussion will show, disciplines are highly complex phenomena; complex to the extent that knowing that something belongs to a certain discipline is not very informative regarding epistemic matters.

What are These Things Called 'Disciplines'?

It is easy to point out difficulties to which any account of disciplines ought to provide some kind of response. Think of questions such as whether philosophy is one discipline in spite of its strong inherent heterogeneity? Or whether statistics is a discipline distinct from the rest of mathematics? Whether all disciplines of the humanities share common characteristics which set them apart relative to disciplines belonging to the natural or social sciences? A general issue is at which level of abstraction distinctions between different disciplines are to be made.

Despite the central importance of one's position regarding issues such as these, it is not uncommon to discuss interdisciplinarity without conveying one's position regarding basic distinctions. Indeed, in most treatments of interdisciplinarity, it is left unclear what exactly is denoted by the term 'discipline'. Neither is it made clear which method is used for distinguishing between disciplines, or why this particular method is preferred.

There are, indeed, many potential ways of creating taxonomies of scientific activities, and there is a substantial body of literature dealing with issues such as what might be the natural (or most useful) way of distinguishing between different branches of science (Bechtel 1986; Darden and Maull 1977; Kellert 2009; Kockelmans 1979a; Sherif 1979). The easiest way forward would be to focus on the organisational structure of scientific institutions, that is, to map how universities are divided into faculties, institutes, departments, units, centres, educations, and so on and so forth. Indeed, this is more or less the "standard method" for deciding whether some interaction is interdisciplinary or not.

It should, however, only take brief reflections to realise that organisational disciplinary boundaries do not necessarily track epistemically relevant differences. It appears that there is nothing to prevent researchers from employing quite similar methods and approaches in spite of being affiliated to different faculties. Consequently, there is an important distinction to be made between the organisational disciplinary divisions on the one hand, and the divisions that would result from analyses of differences in actual scientific approach on the other. The conclusion is, once again, that for an epistemic analysis, one needs a more fine-grained way of distinguishing between scientific activities than the standard method.

Problems are not restricted to cases in which activities considered interdisciplinary are in fact not truly so due to insufficient pre-interaction differences. As Bechtel has stated, if one relies on the most common set of categories, not only is one at risk of considering something to be interdisciplinary in a productive way, when it is not, indeed,

> [...] there may in fact be a great deal of activity that crosses disciplinary boundaries which do not fit the preconception academic and science administrators hold when they bemoan the lack of interdisciplinary activity or of actual scientists when they caution against greater interdisciplinarity. (Bechtel 1986, p. 4)

It is worth pointing out that it is not only important to determine some threshold of sufficient pre-interaction difference, which is indeed a daunting task in itself. There is also the task of figuring out whether some levels of difference or distance between approaches are more likely to lead to good results than others. The following discussion of the nature of disciplines will provide us with elements of a framework for distinguishing relevant activities and differences. This framework has the potential to lead to more accurate and fruitful analyses.

Distinctive Discussions of "Discipline"

As the following will show, the nature of disciplines is a topic that has received quite a lot of philosophical attention. When confronted with the numerous attempts at developing more exact versions of this concept, one might wonder why I want to add another. I shall attempt to provide a reasonable answer to that question below, after going through some of the existing literature on how best to think of disciplines.

The term 'discipline' is often used as if it were a technical term with a clearly explicated sense with which everybody is familiar. On the contrary, however, the word is apparently retrieved more or less directly from ordinary parlance (Kockelmans 1979b, p. 16 f.). It ought not come as a surprise that a vague everyday term does not work very well as the basic unit for accurate analyses of scientific collaborations. Though it is rarely explicitly discussed, this practice has led to a fair amount of confusion. Only rarely do scholars explicitly state that they use 'discipline' in a non-exact sense. The following quotation is an exemplary exception:

> In this volume I talk about disciplines in a broader sense in which they are knowledge-producing enterprises with some shared problems, with some overlapping cognitive tools, and with some shared social structure. Thus, I call philosophy and sociology "disciplines," although there is certainly no set of shared techniques that all practitioners of these fields share. Consider a "Continental" philosopher such as Edward Casey (1997) and an analytic philosopher of science such as John Earman (1989) discussing space, for example. They have little in common in terms of style of argumentation or overlapping references. (Kellert 2009, p. 29; references in the original text)

Problems of Disciplinarity

At first glance, the distribution of disciplines at universities may seem to be organised as a classic taxonomy in which disciplines are grouped together in faculties according to fundamental similarities and divided into more specialised sub-disciplines. Thus, for instance, physics, chemistry, and biology belong to faculties of natural sciences, while literature studies, philosophy, and linguistics belong to faculties of the humanities. Examples of sub-disciplines of physics are quantum mechanics, astronomy, and thermodynamics, while philosophy can be subdivided into, for instance, philosophy of language, philosophy of science, ethics, and so on.

As stated above, upon closer examination it appears that the organisational disciplinary boundaries do not track epistemically relevant differences between scientific approaches (even though they are not completely independent hereof either).

For example, at the University of Copenhagen, psychology is placed at the Faculty of Social Sciences along with sociology, economics, and political science. At the University of Southern Denmark, on the other hand, psychology is placed at the Faculty of Health Sciences along with medicine, public health, and speech and hearing science. First, of course, it is clear that at least one of these universities is not carving academia at its joints. Second, and more interesting, it seems that at least some parts of psychology, for instance empirical and theoretical work on the psychology of concepts, are quite closely related[1] to certain parts of linguistics (especially cognitive semantics) and philosophy of language. Linguistics and philosophy belong to the Faculties of the Humanities at both institutions. From a strictly administrative perspective, thus, collaborations between cognitive semantics and empirical psychology of concepts would be considered interdisciplinary (indeed of the most prestigious cross-faculty kind) at both of these universities. This in spite of the fact that one would be hard pressed to point out significant differences between the preferred methodologies of the involved researchers, and equally hard pressed to point out how the integration of their closely related approaches might result in the development of approaches very different from the original ones.

As another example with a different structure, consider mathematics. Arithmetic and geometry are core parts of any basic curriculum in mathematics. Many problems in geometry are elegantly solved through applications of arithmetic on geometrical issues and vice versa. Indeed, these two elementary parts of mathematics are so interconnected that few would probably think of combinations of arithmetic and geometry as interdisciplinary at all. This impression is strengthened by the observation that groups of people with geometrical and arithmetical expertise tend to overlap significantly.

But indeed, arithmetic and geometry are radically different enterprises targeting completely distinct objects.[2] Indeed, that impressive results could be achieved through the combination of these distinct branches of mathematics was far from obvious to anyone prior to revolutionary work by, among others, Descartes (actually, this continues to be far from obvious to many even today) (Hacking 2014, p. 7). There are other examples of fruitful boundary crossing internal to mathematics such as algebraic geometry, algebraic topology, and analytic number theory. All of these have developed into primary fields of inquiry in their own right. A further, though quite different, example of boundary crossing in mathematics is that of probability and statistics. These have only quite recently come to be parts of mathematics at all, even though by current standards there is no disagreement about where they belong (Galison 2008, p. 115; Gigerenzer et al. 1989). Examples such as these cross-fertilisations between different parts of mathematics[3] speak strongly in favour of hybrid scientific activities. But they speak equally strongly against analysing these activities on the basis of conventional disciplinary distinctions.

Yet a different challenge to the standard understanding of disciplines is the issue of internal incoherence. It is not only philosophy which is plagued by considerable internal heterogeneity. Think of elements of economy such as a priori mathematical modelling, more or less anecdotal descriptions of the history of economic institutions, and highly idealised representations of human beings as strictly rational agents. Do these elements constitute a coherent whole? Or take psychology: What is the connection between intelligence tests, detailed subjective biographies, anonymous lab-tests, factor analysis, and the traumas of childhood? There do not appear to be any common topics or general methodologies involved here (Campbell 1969, p. 332)? Nevertheless, psychology and economics are considered to be coherent disciplines by all traditional standards.

Matters are even less clear with an area such as political science. In political science, work is based on a very heterogeneous multidisciplinary base drawing upon, for example, philosophy, mathematics, history, statistics, sociology, and more (Aldrich 2014, p. 6). One might think of political science as a disciplinary chimera, which could eventually sediment and develop the characteristics of a truly integrated discipline in its own right (whatever these characteristics are). From an epistemic point of view, though, it does seem reasonable to question whether political science has yet developed into a self-contained discipline. And consequently, one might wonder whether a collaboration between political scientists and, say, philosophers should count as interdisciplinary.

On the other hand, though, political science has all the infrastructure needed to be taken seriously as a standalone discipline from an administrative point of view. It has university departments, dedicated journals, peer review structure, PhD-programs, tenure opportunities and so on and so forth. The question remains whether this is sufficient for it to be a discipline in its own right?

Philosophers, historians, and sociologists of science have developed various alternative concepts intended to more fruitfully or adequately capture the basic distinctions between scientific branches. Examples abound: "disciplines" in various versions, "fields", "paradigms", "research programs", "domains", "specialties", "areas", "research groups", and "research networks" (Bechtel 1986; Darden and Maull 1977; Kuhn 1962; Lakatos and Musgrave 1970; Shapere 1974, 1984; Toulmin 1972). Even though few of these were originally developed for the purpose of analysing interdisciplinarity, they draw attention to many issues relevant for the discussion hereof. The following sections will provide the background for offering the alternative concept "approaches".

Three Dimensions (Plus Some) of Disciplinarity

William Bechtel has suggested that different accounts of how to discriminate between disciplines can themselves be characterised by the emphasis they put on three different dimensions: (1) the phenomena they target (i.e. their objects of study); (2) the cognitive tools they use (e.g. the theories, concepts and models they employ); and (3) social structure (e.g. institutional affiliations, peer review organisation, journals) (Bechtel 1986, p. 8).

As the following will demonstrate, Bechtel's framework certainly draws out important aspects of the differences between distinct conceptualisations of the central characteristics of disciplines.

But before we embark on this discussion, the following is worth mentioning: One might reasonably think of Lakatos' "research programmes" and Kuhn's "paradigms" and "disciplinary matrices" as some of the most influential examples of epistemically relevant distinctions between scientific activities. But these concepts are macro-concepts (Hull 1982) and are thus quite far from the level of detail which is required for the present purpose. Therefore, these otherwise important and influential concepts are not included in the following discussion.

Social Aspects Only

One option is to base characterisations of disciplines solely on a single of Bechtel's dimensions. For instance, Nissani (1997) has characterised disciplines as "any comparatively self-contained and isolated domain of human experience which possesses its own community of experts" (p. 203). On Nissani's account of disciplines, social structure is the only aspect considered. Stephen Turner has provided another example based strictly on the social dimension, though he put more stress on educational background. Turner states that "[d]isciplines are kinds of collectivities that include a large proportion of people holding degrees with the same differentiating specialization name [...]" (2000, p. 47). Turner emphasised a number of other (social) aspects which are required for something to be a full-fledged discipline, for example, that the people involved are organised in degree granting units, that these people are employed in these units on the grounds of having obtained such a degree themselves, and that there are other groups who identify themselves with the same discipline name. Turner prided himself for having come up with a definition with a remarkable fit to actual disciplines. This is, however, not much of an achievement. When a definition states little more than that all phenomena called 'x' belong to the category y, it is not surprising if y ends up having a good fit with how the term 'x' is normally used.

Examples discussed above in this chapter have demonstrated that focusing on social aspects of disciplines do not provide a sufficient level of detail for an epistemic analysis of interdisciplinarity. So, let us immediately devote our attention to more auspicious alternatives.

Objects Only

Dudley Shapere's sophisticated notion of scientific *domains* is a good example of an attempt to distinguish between disciplines solely along the dimension of objects of study. There is, indeed, also in this case quite a good fit between the notion of domain and our normal use of the term 'discipline'. The generic discipline biology, for example, corresponds to the study of the very broad domain "living things", whereas the sub-discipline cardiology corresponds to the domain "the human heart".

In Shapere's terminology, a domain is constituted by a set of items, that is, parts that are deliberately selected by the scientist to constitute the object of study. Shapere defined "domain" as the set of items studied in an investigation. Part of studying a domain is to look for relations between items, which indicate that the selected items are fruitfully grouped together (Shapere 1984, p. 279 ff.). Shapere's notion of "domain" is useful in this context, since it brings out important aspects of how to provide a specification of what constitutes the target in a given scientific (representational) activity.

But even though Shapere's account is highly sophisticated, it does not capture disciplinarity adequately. In some cases, activities definitively belonging to the same discipline share few if any objects of study. As an example, few would maintain that Einstein and Newton do not belong to a common discipline, even though their most central objects of study are clearly non-identical. Some would claim they belong to different paradigms, but they are nevertheless both icons of physics.

In the words of Peter Galison:

> Though the objects studied change, of course, there is something roughly commensurate about the nature of physics in 1930, 1950, and 1970. The discipline grew and new specialties arose, but one recognizes across these

years much continuity in the teaching of physics (start with mechanics, proceed to electricity and magnetism, advance to "modern" physics of the atom...). The division into theoretical, mathematical, and experimental physics carries across the mid-twentieth century rather well. True, the boundary between physics and chemistry changes; but (whatever critics said as they raised their anti-Einsteinian voices) no one argued that special relativity was really a piece of chemistry or a fragment of biology—not in 1905, not in 1955, not even in 2005. (Galison 2008, p. 115 f.)

Objects and Tools Combined

Joseph Kockelmans is an example of how to consider disciplines as combinations of the dimensions of objects of study and cognitive tools while excluding considerations of social structure. He defines "discipline" as

> [a] branch of learning or field of study characterized by a body of intersubjectively acceptable knowledge, pertaining to a well-defined realm of entities, systematically established on the basis of generally accepted principles with the help of methodological rules or procedures; e.g. mathematics, chemistry, history. (Kockelmans 1979b, p. 127)

Another example based on the same two dimensions is the discussion of *interfield theories* by Lindley Darden and Nancy Maull (1977), which is a rare example of a philosophical account targeting interdisciplinary collaborations directly. In their interesting discussions of theoretical integration within biology, Darden and Maull also focused strictly on objects of study and cognitive tools while, again, excluding social aspects. As they stated: "We are interested in conceptual, not sociological or institutional, change" (p. 44). Instead of "discipline" they employed the concept "field", which they defined as:

> [...] an area of science consisting of the following elements: a central problem, a domain consisting of items taken to be facts related to that problem, general explanatory factors and goals providing expectations as to how the problem is to be solved, techniques and methods, and, sometimes, but not always, concepts, laws and theories which are related to the problem and

which attempt to realize the explanatory goals. A special vocabulary is often associated with the characteristic elements of a field. (Darden and Maull 1977, p. 44)

A field, in Darden's and Maull's sense, is similar to what one might call a specialised sub-discipline. If biology is a discipline, genetics or cytology would be fields. An interfield theory, then, is a theory (or model) that combines knowledge from two distinct fields in one way or the other.

Darden and Maull listed a number of ways in which interfield theories might improve scientific understanding of some phenomena. As one example, they mentioned how one field may focus on the function of a given phenomenon while another field may focus on its structure. For instance, anatomy is an example of a field, which is distinct from, yet very closely related to physiology. In spite of significant overlaps, anatomy can be distinguished by its study of the structure of the parts which constitute living things, while physiology studies the function of these parts. There is no doubt that in-depth knowledge of the structure of a phenomenon may enrich one's understanding of its function (and vice versa). As another example, one field may be able to specify the physical location of an entity or process which has been postulated within another field. For instance, in the early chromosome theory of Mendelian heredity, genes were postulated to be somewhere in or on the chromosomes. The field of cytology was able to specify that the genes are indeed inside the chromosomes. Thus, collaboration between genetics and cytology helped develop the understanding of this specific part-whole relationship (Darden and Maull 1977, p. 49).

Though Darden and Maull's discussion of fields is quite reminiscent of approach-based analysis, there are substantial differences. One difference worth emphasising is that Darden and Maull focus strictly on issues related to integration internal to biology. Indeed, the integration of fields such as genetics and cytology involves what I would consider the integration of very proximate approaches. But an adequate framework for capturing present day interdisciplinary activities needs a broader scope. Further, whereas the concept "approach" denotes a particular way of representing, the concept "field" is less specific. It lies somewhere between a bundle of approaches and a discipline.

The characterisations proposed by Darden & Maull and by Kockelmans are certainly superior to those based exclusively on social aspects. Nevertheless, an even closer fit can be gained, as the following will show.

All Included?

In his *Human Understanding* (1972) Stephen Toulmin argues in favour of an account of disciplines which include all three of Bechtel's dimensions. But, significantly, Toulmin adds two dimensions (p. 145 ff.). The fourth dimension, which Toulmin places centrally, relates to the problems which the research of a given discipline is aimed at resolving.[4] The fifth dimension is time.

As Toulmin points out, all aspects of disciplinarity targeted in an analysis along the first four dimensions are transitory. The set of people who constitute the members of the discipline and the social structures that bind groups together certainly vary through time. So do the conceptual and theoretical elements (and, for that matter, the instruments used). As emphasised in the Galison quotation just above, even the objects of study sometimes change dramatically in only a few years. And as Toulmin noted, the same is true of the problems addressed. Indeed, on Toulmin's account, changes to the problems addressed is a natural consequence of scientific progress, since he defined scientific problems as *explanatory ideals minus current capacities* (Toulmin 1972, p. 152).

The five dimensional analysis led Toulmin to include atomic physics and law in the category of disciplines, whereas he interestingly excluded philosophy (Ibid., p. 145).

Where Does This Leave Us?

Bechtel's three dimensions serve us well for distinguishing between different ways of categorising science. Indeed, all of Bechtel's dimensions plus Toulmin's two additional ones will have to be included in an adequate characterisation of the basic phenomena of disciplinarity. However, characterising disciplines in this way does not provide us with a very good basis for analysing specific interdisciplinary integrations. I will discuss my preferred alternative in the next section. But first a couple of other remarks are warranted.

All these discussions of the nature of disciplines and how to distinguish between them are certainly interesting. But despite their relevance to the purpose, their conclusions and considerations do not seem to have had much impact on evaluations of research proposals. In such evaluations collaborations are to a very large extent discussed on the basis of institutional affiliations and other social aspects. Little regard is paid to relevant differences or similarities in actual scientific approach.

For one example, one might consult the Guidance for evaluators of Horizon 2020 proposals. Horizon 2020 is a 70 billion € research initiative by the European Union (EC 2014). In the guide, it is made clear that "fostering multi-actor engagement", mixing nationalities, and achieving a higher than average female-to-male ratio among participants are important criteria in the evaluation of research proposals. In specific calls, it is specified more narrowly which kinds of scholars are expected to participate. In very general terms, it is expressed that interdisciplinarity is expected to contribute to the quality and excellence of research proposals. Now, there are many good things to say about mixing nationalities and securing a reasonable male-to-female-ratio. But these criteria certainly do not target epistemic aspects of interdisciplinary activities directly. Apart from very general requirements (objectives must be clear and pertinent; concepts must be sound and novel; the proposed methodology must be credible), no other criteria for evaluating the epistemic aspects of proposals are provided.

This is a quite fundamental problem. Funding (in the real world) is distributed according to criteria such as those quoted above. These criteria display a radically naïve understanding of interdisciplinarity, which is bad news for the future of science, since scientific developments are to a large extent dependent on well-placed external funding.

Part of the problem here, I suspect, is an overreliance on the inadequate distinctions one can make by means of the lay notion of "discipline". As displayed in the above survey, those who have taken the time to consider the issue in detail have not reached agreement. One problem is, I believe, that even some of the subtlest discussions of disciplinarity and interdisciplinarity are carried out using the term 'discipline' as its basis. A more exact or optimal sense of 'discipline' may be developed, of course, but if the term is retained, administrators and policymakers may fail to notice that discussions are no longer based on lay understanding of disciplinarity (if they do consult the specialised literature on the subject at all). This is one potential pitfall, if the language used in specialised literature does not clearly indicate that what is being discussed is something that requires extraordinary interpretative efforts. If alternative terminology is used, readers are forced to notice, at least to some extent, that everyday concepts are considered inadequate for the discussion at hand.

One of the most central conclusions of this book will be that subtle nuances are liable to be lost when elements are transferred between contexts. One might say that concepts, tools, algorithms, vehicles, and, indeed, approaches themselves are often idealised when imported into new settings. This effect is strengthened if the terms used in specialised senses are identical to ones used in everyday senses and no one takes upon him or herself the responsibility of making sure that confusion is kept to a minimum. I will return to this issue in various places that follow.

The attentive reader may at this point be tempted to ask: "But 'approach' seems to be a term retrieved from everyday parlance as well. Is using 'approach' not simply repeating the mistake?" I think not. First, since the everyday sense of approach is much closer to the sense in which I make use of it in the present context,[5] compared to the difference between the everyday sense of 'discipline' and the adequate concept of disciplinarity. Second, the tension involved in claiming that we need to base our analyses of interdisciplinarity on "approach" rather than "discipline" attracts enough attention, I believe, for the reader to give these issues second thought.

It is worth noting the temporal distribution of the different characterisations of "discipline" discussed above. The examples emphasising the epistemic aspects of disciplinarity (i.e. theoretical constructs, targets and methods) are all from the 1960s or 1970s, whereas the discussions focusing primarily on social aspects are from the mid 1980s or later. Presumably, this distribution reflects the impact of the sociological studies of science, which gained their maximum momentum in the slipstream of The Strong Programme in the late 1970s (Barnes 1974, 1977; Bloor 1976).

To some, the above discussion of the nature of disciplines may seem somewhat superficial. However, the point of discussing these philosophical accounts of disciplines is not to come up with yet another even more refined concept. Rather, it is to show that for analysing interdisciplinarity, concepts of disciplines are at the wrong level of abstraction. This does not require discussions in deep philosophical detail of every philosophical position. Of course, one might carry out such a discussion, and it would certainly be interesting. But it is not beneficial to include in this book, I believe.

The bottom line of the discussion of disciplines is, that if the epistemic aspects of scientific hybridisations are to be adequately captured, there is a need for quite a lot more detailed and specific distinctions than what the standard administrative, organisational disciplinary boundaries have to offer. An adequate representation of scientific hybridisations will require a way of zooming in to achieve closer alignment with epistemic differences at a given time. Enter the concept "approach".

Approaches

As a consequence of the considerations about disciplines presented above, I suggest thinking about interdisciplinarity in terms of integration of approaches rather than integration of disciplines.

To repeat, approaches are *specific ways of using a specific vehicle to represent some target*. Importantly, approaches are fixed, invariant entities. They do not evolve or change through time. If you represent differently, you use a different approach.

In any integration, there must be something specific that is transferred, combined, or integrated. Given the complexity of disciplines, it would be very surprising to see someone actually attempting to combine, say, the disciplines of literature studies and neurology in their entirety. However, it is not that unusual to read statements that a given research project is based on interdisciplinary collaboration between disciplines at this quite unspecific level of abstraction.

The above discussion of Bechtel's and Toulmin's frameworks will be useful as we strive for further clarification in the chapters to come:

- First, the social dimension of disciplines seems to be only indirectly relevant to analysis of the epistemic aspects of interdisciplinarity. Indirectly in the sense that we might, perhaps, derive something about what is likely to be involved in an act of integration by focusing on the educational background and affiliations of the people involved. However, there are ways to target the most central epistemic aspects more directly. An approach-based account abstracts from the social dimension. It assumes that it does not matter *who* uses A to represent B. However, it matters *which* A they use and *how* they use it.
- Second, if we are to characterise a discipline such as physics, we certainly need to take temporality into account (as Petr Galison pointed out in the quote above). But the temporal extension of disciplines adds a level of indeterminacy which renders the knowledge that certain disciplines are involved more or less useless in itself. Instead, we need a concept for exactly those temporal cross-sections of the involved disciplines, which serve as the specific input into the interdisciplinary integration in a given case. I suggest that we use "approach" to refer to specific temporal cross-sections of disciplines.
- Third, one might use the same approach to study different phenomena.[6] Thus, at least in some ways, it makes sense to exclude targets from the approach—though targets are not to be completely excluded from analyses of interdisciplinarity, of course.
- Fourth, one might use the same approach for different purposes. Thus, it makes sense to exclude purposes from the approach—though purposes are not to be completely excluded from analyses of interdisciplinarity either, of course.

We are left, then, with vehicles of representation and the tools, algorithms, propositions which are used to constitute the link between vehicle and target. People, targets and purposes are not part of the approach. Therefore, it makes sense to say that an approach is better suited for target A than for target B. Or, for that matter, that an approach is better suited for purpose A than for purpose B.

One might think of approaches as the core which is reapplied to the same type of phenomena in attempts at replication. In such cases, attempts at replication are based on the details included in the published description of the original study. It is completely natural to say, for instance, that when a slightly different approach was used, everything worked as originally expected. Or, that the reason a second group of scientists failed to replicate a given study was that they failed to use the exact same approach.

Studies have examined the extent to which details required for replication are conveyed in psychology papers. Interestingly, this is often not the case (Carp 2012). It is likely that similar problems will haunt many other disciplines potentially involved in interdisciplinarity, and this renders evaluations based on approaches quite difficult. Some of the propositional knowledge that could be made explicit is not included in the text, perhaps because it is part of the taken for granted background assumptions of the people constituting the approach's home discipline. Other aspects may be less easily communicated because they involve skill-like, non-propositional knowledge.

Skills have been treated by scholars in science studies as a central element which is not easily conveyed in publications.[7] The lack of required skills, or lack of attention to the fact that skills are required, are considered to be part of the answer to why some replications fail in some discussions hereof (Collins 1985).

'Skill' is often used to refer to abilities not based on propositional knowledge. Importantly, if a task requires skills, it means that being told how to do something is not sufficient for being able to actually do it. For instance, no one without a considerable set of musical and motor skills can perform *Purple Haze* by Jimi Hendrix on a cello, regardless of whichever elaborate verbal instructions they get. Skills can be acquired, of course (and even sometimes through verbal instruction). Indeed, Collins and Evans consider skills as "something real or fixed that can be transferred from one person to another, or can fail to be transferred" (Collins and Evans 2002, p. 241). Surely, in many cases a skill can be broken down into less mysterious sub-parts, like insight into idealisations or familiarity with a coordinated web of relevant concepts, but also motor abilities acquired through practice and so on.

The analogy between attempts at replication and interdisciplinarity is not perfect, of course. For instance, replication is only relevant in certain types of empirical sciences. The approach notion, in contrast, is intended to capture core aspects of representational activities across all disciplines. What the analogy with replication does suggest, is that an exhaustive analysis of interdisciplinarity in terms of fully explicated pre- and post-interactional approaches is probably impossible. Nevertheless, the descriptions in terms of approaches come quite close to what is actually going on, to the extent that this can be derived from publications. The fully illuminated approach-based analysis, in which all relevant details are revealed, may then serve as a regulative ideal for analyses and evaluations of interdisciplinary activities.

As should be clear by now, the concept "approach" is not intended to replace the concept "discipline". Disciplines (as well as fields, research programmes, and paradigms), will involve a wealth of approaches. Especially when the temporal dimension is taken into account, disciplines will include an abundance of divergent approaches. It may be relevant to some analyses of interdisciplinarity to figure out which approaches are bundled together in specific research projects and programmes.

The discussion of representation in the chapters to follow will enrich the concept of approaches considerably. For now, it will be helpful to revisit the comparison between the concepts "approach" and "field".

Approaches vs. Fields

As quoted above, Darden and Maull make it clear that they are exclusively interested in conceptual and epistemic issues related to the combinations of different fields. In this way, their goals are quite similar to mine. On the other hand, their discussions of interfield theories focus strictly on combinations of very proximate approaches. As they state: "Interfield theories are likely to be generated when two fields share an interest in explaining different aspects of the same phenomenon *and when background knowledge already exists relating the two fields*" (1977, p. 43; my emphasis).

As will be discussed in several places to follow, interdisciplinarity is by no means restricted to collaborations between proximate approaches. When physics is combined with psychiatric diagnosis, or when neuroimaging techniques are combined with literature studies, little existing background knowledge relates the involved approaches prior to their integration.

Indeed, the interfield theories Darden and Maull discuss constitute quite a contrast to the sweeping, unqualified claims of interdisciplinarity involving combinations of very distant approaches. The cases discussed by Darden and Maull are highly successful. But surely, it would be a mistake to argue that interdisciplinarity in general is likely to lead to wonderful results on the basis of a few examples of successful integration of very proximate approaches.

Distance vs. Proximity Revisited

I suggest "measuring" distance between approaches by comparing the assumptions and tools (in a broad sense) used to connect vehicles and targets. The more background assumptions and tools are shared between two approaches, the more proximate they are. Assumptions include things like paradigm cases of good practice, standards of when something is considered sufficiently demonstrated, fundamental stances regarding questions of realism/constructivism, and so forth.

To determine the distance between parent approaches it is, thus, not sufficient to focus on vehicle of representation and target. Two approaches may be quite proximate even when they represent distinct targets by means of distinct vehicles. On the other hand, two approaches may also be quite distant even though they represent the same (or a similar) target by means of the same (or a similar) vehicle.

Construed in this way, the example discussed above of the psychology of concepts and cognitive semantics would involve very proximate approaches. Interestingly, in the discussion above, the collaboration between psychology of concepts and cognitive semantics was considered as "not really" interdisciplinary (in spite of the parent approaches belonging to different faculties) due to the minor differences in their respective approaches. But since then, we have reached the conclusion that there might be great potential for developing fruitful interdisciplinarity in integrations of proximate approaches. So perhaps interdisciplinary collaborations between the psychology of concepts and cognitive semantics are back in business.

A further complication is that the distance between approaches is not only a question of whether the same assumptions and tools are used, but also of whether they are used in the same way. As the discussion of the introduction of operational definitions in psychopathology in Chap. 6 will show, it can make a big difference how an element is put to use. Further, some elements of the intermediate layer may be more important than others. Small variation in one respect may result in two approaches being quite distant, whereas substantial dissimilarities in other respects may not make much of a relevant difference.

This discussion of distance between approaches is highly abstract. The following chapters will provide more concrete exemplification. What should be obvious, though, is that "distance" and "proximity" are not exact, quantifiable measures of the differences between approaches.

There are no a priori reasons to assume that the integration of two proximate approaches will lead to great scientific progress, simply because the approaches are proximate. Neither are there reasons to assume that the integration of distant approaches will lead to poor results, simply because the approaches are distant. But there is likely to be more potential pitfalls to avoid in cases of integration across wide distances, and figuring out whether the result is a good one will require a much more substantial effort.

The more closely related two approaches are, on the other hand, the easier their integration can be expected to be. The ease of integration might also hint in the direction of the novelty of the integrated approach. Ease of integration is partly a consequence of straight up similarities and overlap between the combined approaches, which might indicate that the integrated approach differs to a smaller extent from the parent approaches. But ease of integration is also a result of the involved scientists being already familiar with background assumptions, standard idealisations, and so on.

One might classify the examples discussed by Darden and Maull as micro-interdisciplinarity in the sense that a quite small gap is bridged. This should not, however, be understood as indicating that their examples of proximate integration involve small scientific advances. The examples they analyse certainly constitute genuine and significant steps (if not leaps) forward.

More on Temporality

To make it absolutely clear: As the discussion above shows, not only is there a temporal element to interdisciplinary collaborations, disciplines themselves are temporal entities. Thus, even if one were able to nail down a set of necessary and sufficient criteria for membership of a given discipline at a given point in time, a few years on, these criteria would most likely no longer fit well.

As briefly mentioned above, David Hull pointed out quite a while ago (1982) that concepts such as disciplines, disciplinary matrices, research programs, or research traditions are all macro-concepts constituted by more basic concepts. To take Kuhn's matrices as an example, the constituents would be symbolic generalisations, metaphysical views, models, values, and exemplars (Kuhn 1970, p. 183 ff.). These constituents are all historical entities or continuants, or put in simpler terms, things that change through time. This opens the possibility of thinking of disciplines as evolving entities characterised by continuity through transformations rather than essences. In some cases, the central continuity might be along the dimension of method, in other cases along the dimensions of social structure or topics of interest.

As discussed above, temporality was a central theme in Toulmin's discussion of disciplines. And the following quotation: "If we examine some relatively sophisticated area of science at *a particular stage of its development*, we find that a certain body of information is, *at that stage*, taken to be an object for investigation" (Shapere 1984, p. 273; my emphasis) shows that Dudley Shapere was also well aware of the temporal nature of disciplines.

Several of the most influential philosophers of science (at least since the early 1960s) have had a strong focus on issues related to scientific transformation. Again, Kuhn (1962) is perhaps the best known example, but others are certainly worth noting. Toulmin (1972) and Laudan (1977) both allow for more or less complete transformations within a discipline, as long as change is gradual. Lakatos (1970) allow for change in less central parts of the system, but require the retention of some theoretical "hard core".

One consequence of viewing disciplines in these ways is that there is nothing in principle to hinder two disciplines from evolving into using very similar methods for solving identical problems—even if they are organisationally unrelated. Or for that matter, though they might share common ancestry, two disciplines may evolve into using completely distinct methods for solving distinct problems.

In light of the above, interdisciplinarity can be considered as sudden, more or less radical, transformations as compared to the slow, incremental developments of intra-disciplinary scientific evolution. According to Toulmin, gradual developments in explanatory aspirations and capacities are central to rational enterprises. But that would deem the explorative conceptual leaps of many interdisciplinary projects irrational.

It is central to my framework that analysis of interdisciplinary integrations should be based on temporal cross-sections of the involved disciplines. Apart from a gain in specificity, this also highlights that future developments of the parent disciplines are not automatically built into the generated hybrid. Put differently, the hybrid account will not automatically update in accordance with developments in one or more of its parent disciplines. This is a central point in Patricia Kitcher's analysis of the failures of psychoanalysis (1992, 2007). Below, I will provide some further examples of how serious this problem can be.

Summing Up

Based on the above discussion, it is clear that even though the concept "discipline" is certainly not without utility, something more specific is required for an epistemic analysis of interdisciplinary activities. An approach, on the other hand, is defined as a snapshot of a specific way of representing certain phenomena at a given point in the development of a discipline. It is thus a fixed, invariant phenomenon which does not change through time. That someone used a given approach at a given time does not change, because they use a different approach at some other time. While disciplines are constantly evolving, approaches remain fixed.

For an exact analysis of interdisciplinary activities to be possible, we need to get a hold of what exactly is combined. Any interdisciplinary activity must result in some product (unless of course it is abandoned, for some reason, and does not lead to any results worth communicating). This product will incorporate some approach to the representation of some target. All the elements used must be picked from somewhere unless they are freshly developed (e.g. in the relatively rare case where a completely original statistical tool or propositional algorithm or a hitherto unheard of basic assumption is put to use). It must therefore be possible to single out the specific approach from where a certain element is imported (it may of course be used in several places). In many situations, the lineage may be indeterminable to some extent. But such lack of clarity is worth spelling out in its own right, since it is likely to require extra attention if one is to evaluate the adequacy of the hybrid approach.

Sometimes, of course, it might only be fragments of approaches which are combined. In such cases, it might also be illuminating to consider the function of the combined fragments in the parent approaches, and compare this function to how they are put to use in the integrated approach.

Notes

1. For instance, in terms of topics of interest, central aspects of methodology, paradigmatic examples of good practice, as well as theoretical background assumptions (Fauconnier and Turner 2002; Lakoff 1987; Margolis and Laurence 1999; Murphy 2002).
2. Just think of how Kant considered arithmetic to be the synthetic a priori truths of time, while he considered geometry to be the synthetic a priori truths of space [4:283].

3. —which is "a constant source of both delight and achievement among mathematicians" (Hacking 2014, p. 9).
4. As explicitly stated in the quotation above, Darden and Maul do indeed include problems in their characteristic of fields, as well.
5. According to www.oxforddictionaries.com, an approach is "a way of dealing with a situation or problem". In the present context, of course, focus is on representational approaches, and the situation or problem is how to scientifically represent some phenomena. You can decide to adopt an entirely different approach or to slightly alter your present approach. Pretty straightforward, in my opinion.
6. There are some difficulties related to applying the exact same approach to different targets, though. These issues will be addressed in Chap. 7.
7. Indeed, grasping the concept "skill" may require some sort of meta-skill. In the words of Collins and Evans: "[S]kills [are] notoriously hard to explain—as qualitative sociologists know to their cost" (2002, p. 258).

References

Aldrich, John H. 2014. *Interdisciplinarity*. New York: OUP.

Barnes, Barry. 1974. *Scientific Knowledge and Sociological Theory*. London: Routledge and Kegan Paul.

———. 1977. *Interests and the Growth of Knowledge*. London: Routledge and Kegan Paul.

Bechtel, William, ed. 1986. *Integrating Scientific Disciplines*. Dordrecht: Martinus Nijhoff.

Bloor, David. 1976. *Knowledge and Social Imagery*. London: Routledge and Kegan Paul.

Campbell, Donald T. 1969. Ethnocentrism of Disciplines and the Fish-Scale Model of Omniscience. In *Interdisciplinary Relationships in the Social Sciences*, ed. Muzafer Sherif and Carolyn W. Sherif. Chicago: Aldine Publishing Company.

Carp, Joshua. 2012. The Secret Lives of Experiments: Methods Reporting in the fMRI Literature. *NeuroImage* 63: 289–300.

Collins, Harry M. 1985. *Changing Order: Replication and Induction in Scientific Practice*. London: The University of Chicago Press.

Collins, Harry M., and Robert Evans. 2002. The Third Wave of Science Studies: Studies of Expertise and Experience. *Social Studies of Science* 32 (2): 235–296.

Darden, Lindley, and Nancy Maull. 1977. Interfield Theories. *Philosophy of Science* 44 (1): 43–64.

EC. 2014. Guidance for Evaluators of Horizon 2020 Proposals. [pdf]. EC. Last modified 26 September 2014. Accessed 8 Januray. http://ec.europa.eu/research/participants/data/ref/h2020/grants_manual/pse/h2020-evaluation-faq_en.pdf.

Fauconnier, Gilles, and Mark Turner. 2002. *The Way We Think: Conceptual Blending and the Mind's Hidden Complexities*. New York; Great Britain: Basic Books.

Galison, Peter. 2008. Ten Problems in History and Philosophy of Science. *Isis* 99 (1): 111–124.

Gigerenzer, Gerd, Zeno Swijtink, Theodore Porter, Lorraine Daston, John Beatty, and Lorenz Kruger. 1989. *The Empire of Chance: How Probability Changed Science and Everyday Life (Ideas in Context)*. Vol. 12. Cambridge: Cambridge University Press.

Hacking, Ian. 2014. *Why is There Philosophy of Mathematics at All?* Cambridge: Cambridge University Press.

Hull, David L. 1982. Exemplars and Scientific Change. In *Proceedings of the Biennial Meeting of the Philosophy of Science Association*. Chicago, IL: University of Chicago Press.

Kellert, Stephen H. 2009. *Borrowed Knowledge: Chaos Theory and the Challange of Learning Across Desciplines*. Chicago, IL: University of Chicago Press.

Kitcher, Patricia. 1992. *Freud's Dream: A Complete Interdisciplinary Science of Mind*. Cambridge, MA; London: MIT Press.

———. 2007. Freud's Interdisciplinary Fiasco. In *The Prehistory of Cognitive Science*, ed. Andrew Brook, 230–249. Basingstoke, UK; New York: Palgrave Macmillan.

Kockelmans, Joseph J., ed. 1979a. *Interdisciplinarity and Higher Education*. Pennsylvania: The Pennsylvania State University Press.

———. 1979b. Science and Discipline: Some Historical and Critical Reflections. In *Interdisciplinarity and Higher Education*, ed. Joseph J. Kockelmans. Pennsylvania: The Pennsylvania State University Press.

Kuhn, Thomas S. 1962. *The Structure of Scientific Revolutions*. International Encyclopedia of Unified Science: Foundations of the Unity of Science V. 2, No. 2. Chicago: University of Chicago Press.

———. 1970. *The Structure of Scientific Revolutions*. International Encyclopedia of Unified Science. Foundations of the Unity of Science, V. 2, No. 2. 2nd ed. Chicago: University of Chicago Press.

Lakatos, Imre. 1970. Falsification and the Methodology of Scientific Research Programmes. In *Criticism and the Growth of Knowledge*, ed. Imre Lakatos and Alan Musgrave, 91–196. Cambridge: Cambridge University Press.

Lakatos, Imre, and Alan Musgrave. 1970. *Criticism and the Growth of Knowledge: Proceedings of the International Colloquium in the Philosophy of Science*. Cambridge: Cambridge University Press.

Lakoff, George. 1987. *Women, Fire, and Dangerous Things: What Categories Reveal About the Mind*. Chicago; London: University of Chicago Press.

Laudan, Larry. 1977. *Progress and Its Problems: Toward a Theory of Scientific Growth*. Berkeley: University of California Press.

Margolis, Eric, and Stephen Laurence. 1999. *Concepts: Core Readings*. Cambridge, MA; London: The MIT Press.

Murphy, Gregory L. 2002. *The Big Book of Concepts*. Cambridge, MA; London: MIT Press.

Nissani, Moti. 1997. Ten Cheers for Interdisciplinarity: The Case for Interdisciplinary Knowledge and Research. *Social Science Journal* 34 (2): 201–216.

Shapere, Dudley. 1974. *Galileo: A Philosophical Study*. Chicago, IL: University of Chicago Press.

———. 1984. *Reason and the Search for Knowledge: Investigations in the Philosophy of Science, Boston Studies in the Philosophy of Science*. Dordrecht: Reidel Publishing.

Sherif, Muzafer. 1979. Crossdisciplinary Coordination in the Social Sciences. In *Interdisciplinarity and Higher Education*, ed. Joseph J. Kockelmans. Pennsylvania: The Pennsylvania State University Press.

Toulmin, Stephen. 1972. *Human Understanding*. Oxford: Clarendon.

Turner, Stephen. 2000. What are Disciplines? And How is Interdisciplinarity Different? In *Practising Interdisciplinarity*, ed. Peter Weingart and Nico Stehr. Toronto: University of Toronto Press.

3

Interdisciplinarity Studies

With this chapter, I provide a somewhat selective survey of existing approaches to the study of interdisciplinarity as well as representative examples of interdisciplinary scientific activities. The account is intended to illustrate a very central part of the motivation for developing an alternative approach to the analysis of interdisciplinarity, namely that existing treatments are analytically biased. The bias leads to a strong emphasis of social aspects of interdisciplinary collaborations. This emphasis has resulted in an unfortunate misrepresentation of interdisciplinary science in the Interdisciplinarity Studies literature.

This chapter will also point out some of the general trends in recent interdisciplinary research as well as noteworthy historical examples. These will show, first, that interdisciplinarity is no recent invention, and, second, that there are certainly good reasons for drawing inspiration from examples of successful integration. These successes cannot fully explain, though, why interdisciplinarity has evolved into such an attractive way of developing science.

Following Shalinsky (1989) I use "polydisciplinary" (and the derivative "polydisciplinarity") as a generic concept subsuming all the different types of scientific collaboration discussed in this chapter. In the rest of the book, I will focus strictly on interdisciplinarity, the subcategory of

© The Author(s) 2018
R. Hvidtfeldt, *The Structure of Interdisciplinary Science*, New Directions in the Philosophy of Science, https://doi.org/10.1007/978-3-319-90872-4_3

polydisciplinarity in which disciplinary inputs are allegedly integrated (somehow).

The academic literature on interdisciplinarity is large and packed with thorough and interesting studies of various aspects of polydisciplinary collaborations. Nevertheless, the treatment that the topic of polydisciplinarity has typically received within Interdisciplinarity Studies has contributed to the generation of a widespread and unfortunate tendency which, polemically put, considers interdisciplinarity to be "inherently good". As a consequence, far from sufficient attention is paid to the special epistemic pitfalls involved in interdisciplinary practice.

Hopefully, the arguments of this book may help to direct attention to the fact that, in some cases at least, serious epistemic difficulties do arise when scientific approaches are integrated.

This chapter will not provide an exhaustive survey of Interdisciplinarity Studies, or, for that matter, provide necessary and sufficient criteria for counting as part of this movement. It will suffice for present purposes to account for the positions of some of the most prominent and influential names as presented in some of the most prototypical publications.[1] Neither shall I ascribe the attitude I argue against, that is, the analytical attitude that results from an overemphasis of social aspects of interdisciplinarity, to any particular person or group of persons within Interdisciplinarity Studies or elsewhere. It does not matter for the present concern whether anyone would openly adhere to this position. The important issue is that central writings on the topic of interdisciplinarity showcase this bias. Indeed, I believe that most readers familiar with scientific practice in, especially, Europe and North America will recognise the general attitude I am targeting—not least if the reader is acquainted with science policy and research funding.

A small number of philosophers work on analysing aspects of interdisciplinarity, some of them somewhat along the lines of the account developed in this book.[2] So far, however, no general framework for analysing interdisciplinarity has been developed. Rather, existing treatments focus on detailed analyses of specific instances of interdisciplinary collaboration. Much of this work is of splendid quality, though, and some of it will be discussed in the present chapter, in order to draw and build on those efforts in the following chapters.

The positive account offered in the chapters that follow is by far the most interesting part of this book. The partly critical, partly historical account in this chapter may be less interesting in itself. But it serves the important purpose of pointing out the cavity which the subsequent positive account is intended to fill. On this background, most people interested in the topic of interdisciplinarity will hopefully sympathise with the position that epistemic aspects of polydisciplinary activities have so far received less attention than they deserve.

What is Interdisciplinarity?

There seems to be widespread agreement that there are no commonly accepted definitions of terms such as 'crossdisciplinarity', 'pluridisciplinarity', 'multidisciplinarity', 'interdisciplinarity', or 'transdisciplinarity'. All of those terms are considered to denote variations of polydisciplinary activities—that is, activities which draw on and combine (in some way) elements from at least two distinct disciplinary contexts.[3] Apart from 'polydisciplinarity', all those terms were parts of an influential taxonomy developed by Erich Jantsch (Jantsch 1970; Klein 2010). Despite disagreements on the exact sense of the term, Erich Jantsch is generally credited for having invested 'interdisciplinarity' with its present meaning.

Quite a lot of effort has been put into developing various taxonomies of scientific collaboration. Fortunately, for the present discussion there is no need to stipulate exact definitions of any of the above notions, let alone attempt to resolve the widespread disagreements regarding their exact meaning. For present purposes, the approximate outlines of the meanings of the terms provided in the next few paragraphs will suffice.

One very fundamental issue, on which there seems to be agreement, is that interdisciplinarity is characterised by *integration* in one form or another. Indeed, this is often considered the litmus test of interdisciplinarity as compared to, most significantly, multidisciplinarity (Klein 2010; Lattuca 2001, pp. 78, 109).

'*Multidisciplinarity*' is widely used to denote collaborations in which two or more scientific disciplines are "placed side by side" (perhaps in an effort aimed at solving a common problem). Importantly, this is done in a way that does not result in the development of hybrid or novel knowledge, and disciplinary identities are not called into question.

Julie Thompson Klein considers multidisciplinarity to be

> [...] an approach that juxtaposes disciplines. Juxtaposition fosters wider knowledge, information, and methods. Yet, disciplines remain separate, disciplinary elements retain their original identity, and the existing structure of knowledge is not questioned. This tendency is evident in conferences, publications, and research projects that present different views of the same topic or problem in serial order. Similarly, many so-called 'interdisciplinary' curricula are actually a multidisciplinary assemblage of disciplinary courses. [...] In [such cases] [...] integration and interaction are lacking. (Klein 2010, p. 17)

Some polydisciplinary activities are, thus, interdisciplinary by name only. And the difference between these and the truly interdisciplinary ones, is the lack of integration and interaction in the former.

In Jantsch's original taxonomy, '*transdisciplinarity*' was used to refer to the all-encompassing "coordination of all disciplines and interdisciplines in the education/innovation system on the basis of a generalised axiomatics" developed for social purposes (Jantsch 1970, p. 411). Most contemporary uses of the term are less ambitious and only require that collaborations somehow transcend academia. Transcendence can either be achieved by developing an overarching scientific synthesis or by involving sectors external to science such as the private sector or simply non-academic members of some relevant community (Klein 2010, p. 24 f.).

'*Crossdisciplinarity*' and '*pluridisciplinarity*' are less frequently used terms. The latter was originally thought by Jantsch to indicate a version of multidisciplinarity in which the involved disciplines were less isolated from each other. This would somehow facilitate the building of relationships between the involved disciplines. 'Crossdisciplinarity', on the other hand, denoted situations in which aspects of one discipline are somehow imposed upon other disciplines in a more or less imperialistic manner[4] (Jantsch 1970, p. 411). In other contexts, though, 'crossdisciplinarity' is used as a generic term for all types of collaboration involving several disciplines including the above mentioned ones (e.g. Bechtel 1986, p. 22).

Thus understood, 'crossdisciplinarity' is synonymous with 'polydisciplinarity' in the sense stipulated above.

Knowledge Generation and Integration

To repeat, the most important aspect of the above account of subtypes of polydisciplinarity is that integration and interaction are considered to be the central characteristics that set interdisciplinary activities apart from the other sub-types. On many occasions, Klein has clearly stated that this is her conviction. As a few examples:

> When integration and interaction becomes proactive, the line between multidisciplinarity and interdisciplinarity is crossed. (Klein 1990, p. 18)

And:

> [T]he only true interdisciplinarity is *integrated interdisciplinarity*, which [can be described as] collaborations in which the concepts and insights of one discipline contribute to the problems and theories of another. (Klein 1990, p. 20)

Also central, however, is that the integration leads to the generation of (new/integrated/mutual) knowledge:

> Mutual knowledge emerges as novel insights are generated, disciplinary relationships redefined, and integrative frameworks built. (Klein 2008, p. 119)

So, knowledge and knowledge generation, however one likes to think of it, is certainly considered to be central to interdisciplinarity. Indeed, the last sentence in the foreword of *The Oxford Handbook of Interdisciplinarity* is the following:

> [...] we hope that this handbook can contribute to critical assessment of a vibrant new dimension of knowledge production. (Frodeman et al. 2010, p. iv)

Notice how "knowledge production", the emergence of "mutual knowledge", the generation of "novel insights", and that "concepts and insights of one discipline contribute to the problems and theories of another" are considered to be essentially different from the fostering of "wider knowledge, information, and methods" in multidisciplinary activities. Interdisciplinarity results in *new knowledge*, somehow—not just deeper, more inclusive, or more wide-reaching knowledge. Unfortunately, the processes involved in generating "new knowledge" as opposed to "wider knowledge" are not discussed in any detail. Consequently, insufficient attention is paid to the specific epistemic processes apparently central to interdisciplinary activities.

In the existing literature on interdisciplinarity it is, indeed, sometimes questioned whether the production of new knowledge is necessarily a good thing. In the introduction to *The Oxford Handbook of Interdisciplinarity* it is even stated that "[…] it is evident that knowledge can sometimes do more harm than good" (Frodeman et al. 2010, p. xxix). It is then concluded that the remedy against producing too much knowledge is to gain knowledge, which enables one to decide whether knowledge in a given situation is pertinent or not (p. xxx). How one can know the value of knowing something before gaining the knowledge which one is evaluating remains somewhat mysterious, though.

The arguments presented in this book constitute an attempt at raising the level of attention to the dynamics of interdisciplinarity in the sense of integrative polydisciplinarity. The primary focus of the methodology developed below will be on how to analyse integration, and what the analysis of integration might tell us about interdisciplinarity and representation in general. Showing that integration of approaches may involve substantial epistemic difficulties will constitute a strong argument in favour of careful conduct in cases of interdisciplinary science.

Even though I will focus strictly on interdisciplinarity, let me note that I do indeed find discussions of other versions of polydisciplinarity important—especially the (potential) activities picked out by the concept "transdisciplinarity" (especially in the less ambitious sense discussed above). It seems reasonable, however, to attempt to get a hold on what is going on in the "simpler" interdisciplinary settings before attempting to comprehend the even more complicated dynamics of transdisciplinarity.

It is my conviction, though, that the discussions of interdisciplinarity of this book can be extended to capture transdisciplinary collaborations as well. But we need to get the basic dynamics of the simpler (though by no means simple) cases worked out first.

Interdisciplinarity is Not New

That drawing inspiration from other scientists is a central part of developing new and improved scientific approaches may seem almost too obvious to discuss. But in this context, it is at least worth mentioning that drawing inspiration from more than one distinct scientific approach and attempting to develop integrated approaches hereof has been a central part of science for a very long time.

As Stephen H. Kellert writes in his *Borrowing Knowledge* (2009):

> Economist Herbert Simon pointed out in 1959 that "the social sciences have been accustomed to look for models in the most spectacular of the natural sciences," and he went on to say "there is no harm in that, provided that it is not done in a spirit of slavish imitation" [...]. Knowledge has always traveled between disciplines, from Darwin's use of geological facts to the role of radioactive dating in archaeology to contemporary discussions of historical "forces" and social "inertia" [...]. The insights of linear Newtonian physics have sometimes proven useful for conceptualizing human social change. (p. 13)

One might be tempted to point out that Kellert's use of 'always' appears somewhat excessive, if the period he refers to only goes as far back as the works of Charles Darwin.[5] Quibbles aside, however, it is not difficult to find earlier examples of borrowed knowledge. A good example is David Hume's *A Treatise of Human Nature* (1738) with the telling subtitle *Being an Attempt to Introduce the Experimental Method of Reasoning into Moral Subjects*. In its introduction, Hume described the idea of "the application of experimental philosophy to moral subjects" and how "some late philosophers in England, [...] have begun to put the science of man on a new footing, and have [thereby] engaged the attention, and excited the curiosity of the public" (Hume 1738, p. 26).

Hume's treatise is a clear example of an attempt at integrating distinct areas of inquiry, i.e. the study of moral subjects with experimental philosophy.[6] And not only is the idea of integrating distinct areas not new, neither, apparently, is its potential to attract public interest. Hume continued:

> [...] [T]o me it seems evident, that the essence of the mind being equally unknown to us with that of external bodies, it must be equally impossible to form any notion of its powers and qualities otherwise than from careful and exact experiments, and the observation of those particular effects, which result from its different circumstances and situations. And though we must endeavour to render all our principles as universal as possible, by tracing up our experiments to the utmost, and explaining all effects from the simplest and fewest causes, it is still certain we cannot go beyond experience [...]. (Hume 1738, p. 27 f.)

If interdisciplinarity is thought of in terms of developing and improving science by having "the concepts and insights of one discipline contribute to the problems and theories of another", Hume's treatise certainly qualifies.

In (2014) Simone Mammola traced attempts at, or at least belief in the potential benefits from, integration between medicine and philosophy as far back as Aristotle's discussions hereof in the collection of texts referred to as *Parva Naturalia*. Then, at least, it appears reasonable to talk of 'always' with respect to the topic of science.[7]

As a final, more recent, example, think of *The Structure of Scientific Revolutions* by Thomas Kuhn (1962); the well-known opening lines of which read:

> History, if viewed as a repository for more than anecdote or chronology, could produce a decisive transformation in the image of science by which we are now possessed. (Kuhn 1962)

It is common knowledge that Kuhn did in fact bring about decisive transformations within philosophy of science, as well as in the general understanding of science, by developing a historically informed approach to the study of science. Kuhn is especially interesting in that his work is one of the main inspirations for the social turn in philosophy of science as well as a paradigm example of quite successful disciplinary integration. Certainly, Kuhn could be claimed to have developed "new knowledge", or perhaps a new way of knowing something about science.

The above examples are only a handful out of the many interesting historical attempts at integrating (elements from) distinct disciplines. Though far from exhaustive, the above account suffices to show that interdisciplinarity is no recent invention.

What is indeed a more recent invention, is the tendency to consider interdisciplinarity as the heart of scientific innovation—as something that is more or less necessary for driving scientific innovation forward. In the next part of this chapter, I will attempt to diagnose these recent and dominant trends in western science.

More Recent Developments: The Turn-Turn

In contemporary discussions of science, the ability to bring about paradigm shifts is a central characteristic of true scientific genius. As a result, the influence of Kuhn may have motivated many scientists to look for ways to bring about paradigm shifts themselves, or, more modestly, at least a "turn" in the way we view some matter. One might say that since Kuhn there has been a "turn-turn" in approaches to developing science.

Kuhn himself set an example with his historical turn in philosophy of science, which, as mentioned, in turn inspired the social turn in popular approaches to studying science. More recently we have witnessed the cognitive turn affecting all kinds of disciplines, much like the evolutionary turn and the neurological turn. Nowadays, disciplines twist and turn so frequently that the achievements of one turn hardly have time to sediment before the next wave of revolutionary rotation sets in.

Interdisciplinarity is essentially a way of turning things upside down. It is not easy to measure exactly the influence of Erich Jantsch and Interdisciplinarity Studies on the development of the present widespread enthusiasm for interdisciplinarity. But it is not unfair to claim that inter-disciplinarity has developed into being, if not a goal in itself, then to a large extent a prerequisite for scientific projects to be considered innovative and well-formed.

In recent decades, a tendency has developed of drawing inspiration from especially successful or high-profiled scientific branches. Especially aspects from neurology and evolutionary theory have been imported into more or less all other kinds of settings—from art studies and musicology to psychopathology and the behavioural sciences.

There are different ways to close in on the multitude of attempts at creating interdisciplinary hybrids in contemporary science. The search for grand scale turns is one strategy. Another is to focus on a single discipline and look for which kinds of other disciplinary influences have been imported, in order to develop novel approaches. As one example of this perspective, one may find many recently developed approaches to the analysis of literature which are explicitly interdisciplinary.

Literature Studies

Think of examples such as *Cognitive literature studies* in which cognitive neuroscience, cognitive psychology, and philosophy of mind (among other things) are applied to the study of literature and other aspects of culture traditionally belonging to the humanities (Aldama 2010; Crane 2000; Crane and Richardson 1999; Zunshine 2015). Or think of Darwinian art studies in which literature and art is situated in the context of evolution and natural selection. Literature and art (and aesthetically related abilities in general) can thus be interpreted as adaptive in one way or the other. This can lead to new interpretations of works of art themselves, or of the role of art in human survival and selective processes.

For instance, Brian Boyd has argued that the ability to create stories is adaptive since it provides humans with a "free space" for developing innovative ideas and spreading the idea of innovation in the first place:

> Stories, whether true or false, appeal to our interest in others, but fiction can especially appeal by inventing events with an intensity and surprise that fact rarely permits. Fictions foster *cooperation* by engaging and attuning our social and moral emotions and values, and *creativity* by enticing us

to think beyond the immediate in the way our minds are most naturally disposed—in terms of social actions. (Boyd 2009, p. 382 f.; emphasis in the original)

According to Boyd, stories further help humans develop communal social identity, which he also considers adaptive.

As related examples, in (1997) Steven Pinker argued that humans have evolved the abilities to appreciate and produce literary narratives since these may provide data relevant for adaptive issues. Several other authors argue that aesthetic responsiveness is adaptive since it helps organise impressions of reality by means of emotionally and aesthetically modulated cognitive models (Carroll 2004; Dissanayake 2000; Dutton 2009). All of these approaches focus on the benefits of the discussed abilities with respect to survival.

Choosing a different emphasis, Geoffrey Miller has argued that sexual selection is the driving force behind the development of abilities such as humour, wit, musicality, and the ability to produce lyrical and exciting prose. People with those kinds of talents are more sexually attractive,[8] it is argued, and therefore these traits have been selected for. In line with Darwin (1871) and Fisher (1930) among others, Miller distinguishes sharply between *natural selection* (which is related to traits that increase overall fitness, e.g. with respect to escaping predators or locating food) and *sexual selection*, which increases chances of producing more offspring by intensifying sexual attractiveness. Miller, thus, likens aesthetic ability to the paradigm example of a sexually selected feature: The tail of the peacock. The peacock's tail may impress licentious peahens, while it actually seems to diminish fitness in terms of the ability to hide or escape from, say, hungry tigers.

Psychoanalytic Literature Studies

Take another example from the study of literature: Drawing on examples like Sigmund Freud, Jacques Lacan, Slavoj Žižek, and others, psychoanalytic literary criticism focuses on the role of consciousnesses and the unconscious in literature. Objects of analysis include the author, the reader, and the (fictive) characters of the text.

In such analyses, literary works are interpreted as expressions of, for instance, repressed emotional and psychological conflicts. A central assumption is that, for instance, the actions of characters in a novel somehow display the traumatic events of the author's childhood, sexual experiences, and family life as well as neuroses and fixations from which he or she may suffer. All of this is only expressed indirectly and metaphorically in the text. Analytical techniques inspired by central Freudian concepts such as "symbolism", "condensation", and "displacement" are therefore applied to unveil the real content (which is not identical to the authors' intention with the text).

Psychoanalysis is an interesting example, since it is, to say the least, a disputed framework within its parent discipline of psychiatry. And the theoretical background, on which many of the central assumptions of psychoanalysis rest, have been thoroughly, and more or less unanimously, rejected. For instance, beliefs in recapitulationism or Lamarckian inheritance have been out of academic favour for quite a while. But both of these elements are indispensable for Freud's reasoning (Kitcher 1992, 2007; Rasmussen 1991). I do not mean to pass judgment on psychoanalysis, but it is an interesting question to what extent the considerable difficulties facing the original approach are taken into account when psychoanalysis is put to new use in literature studies? The further, and no less interesting, question is whether it is reasonable or fruitful to base an approach in, say, literature studies on a parent approach which is a manifest failure?[9]

In my opinion, one should not expect general conclusions regarding issues such as these. Let us call this 'insight B': We should not assume that some element will not work well in a new context, simply because it did not work well in its context of origin.[10]

A Different Approach

These were only a few, though notable, examples. There are many others equally worthy of discussion. However, let us turn things upside down and look at the source disciplines from which inspirations are drawn. Doing so will reveal a different, but no less interesting, pattern of contemporary interdisciplinarity. There are a number of trendy sources for providing input to other disciplines. The most notable are, perhaps, the already mentioned scientific fields of evolutionary theory, neurology, and cognitive science.[11]

Some of these sources have been widely applied outside of aesthetics. Take cognitive science as a first example. All kinds of disciplines come in a cognitive variety. Some of these, such as cognitive linguistics and cognitive psychology, have certainly been tremendously successful and have had wide applications outside of science such as in language teaching in pre-schools and kindergartens as well as in therapeutic approaches within clinical psychology.[12]

One central element, which has contributed to the success of therapeutic cognitive psychology, has been a focus on proceeding in the so-called "evidence-based" manner, which puts emphasis on quantifying and documenting one's efforts and results. Since other approaches to psychotherapy (psychodynamic therapy, for instance) have been less good at documenting their results, cognitive therapy has come to appear as much more successful on this basis alone (Shedler 2010; Wampold et al. 2011).

The Evolutionary Turn

Evolutionary theory has also been applied to a tremendous diversity of topics in recent decades. A few examples from literature and art studies were discussed above. There is an abundance of interesting applications of evolutionary theory to countless other disciplines, though. Among the many examples are evolutionary anthropology, evolutionary ecology, evolutionary neuroscience, evolutionary psychology, and, indeed, evolutionary epistemology (one version of which might be more precisely termed 'evolutionary philosophy of science').

In psychopathology, all kinds of applications of evolutionary theory have been constructed, for instance in attempts at explaining why crippling mental disorders have not been selected against. In some cases, mental disorders are considered as something that have been adaptive in some EEA[13] but have ceased to be beneficial in modern societies.

Over the past 10,000 years, roughly since the invention of agriculture, humans have so transformed their way of life that for the majority of the world's six billion people, conditions today are now vastly different to those in the EEA. Could this transformation be the root cause of several psychiatric disorders? This view, sometimes called the 'out of Eden' hypothesis, 'genome lag' or the mismatch hypothesis, has attracted numerous adherents. (Cartwright 2007, p. 299 f.)

The specific phobias, such as fear of snakes or spiders, are among most plausible candidates for something that might have been adaptive in the EEA, but is less advantageous today. In a large part of the civilised world, especially in urban areas, fearing snakes and spiders is a waste of energy, but things might have looked differently in the EEA of homo sapiens.[14] Many other out-of-eden-style explanations for mental disorders are considerably less plausible, though.

One such example focuses on schizophrenia, which is interpreted as having had a group-splitting function in the EEA. A lot of evidence indicates that schizophrenia has a strong genetic component. Other evidence points out that schizophrenia is no recent phenomenon. There is approximately a 1% risk of schizophrenia across all cultures, even ones which were isolated long ago, such as Australian aboriginals which, on some accounts, split off about 60.000 years ago. This indicates that schizophrenia was genetically well-established before then.

But since schizophrenia is a very debilitating condition, why has it not been selected against? Compared to the average person, people suffering from schizophrenia produce less offspring and die relatively young even if they do not commit suicide (which they quite often do). In almost any respect schizophrenia immediately comes across as a severe disadvantage to the individual.

John S. Price and Anthony Stevens (e.g. 1999) have suggested that in the EEA there were drawbacks related to groups increasing beyond a certain size. For instance, it might be difficult to gather enough food locally to feed a very large group. To avoid devastating social tensions, it would then be adaptive if some group-splitting function set in when groups reached critical sizes. Since the prevalence of schizophrenia is 1%, the likelihood of one member suffering from schizophrenia would be high when group size approached one hundred. If schizophrenia served a group splitting function, two groups of more optimal sizes would result.

Stevens and Price argue that the individuals showing symptoms of schizophrenia might somehow have attracted followers who were discontent with their present situation in the (too) large group. They base this in part on an assumption that well-known charismatic leaders (such as Adolf Hitler) might have suffered from schizophrenia.

In support of their hypothesis, Price and Stevens add that

> [...] the schizophrenic individual is behaving like a prophet lacking followers. The prophet is alienated from normal social intercourse because of his deviant beliefs, and in the absence of followers, his preaching of his mission having failed, he might be expected to withdraw from society and to refrain from social intercourse. (Price and Stevens 1999, p. 201)

This is tremendously speculative. Where are the references to empirical studies of how prophets usually act when deprived of followers, for instance? One might also worry about the propensity for feeling fearful or disgusted that many people exhibit when faced with apparent insanity. We need some explanation for why we should not expect this reaction to have occurred in the alleged EEA. Taking such reactions into account makes it seem more likely that individuals suffering from schizophrenia would be expelled or even killed, than that they would end up as charismatic leaders of newly formed groups.

Indeed, it is not uncommon to encounter tensions between the reality of people suffering from some condition and the function evolutionary theorists assume they might have fulfilled in the EEA. As Dominic Murphy has pointed out, in many cases of evolutionary theories of mental disorders there is a "failure [...] of psychological form not matching alleged biological function". (2005, p. 761). In plain terms: Very few people suffering from schizophrenia are plausible candidates for charismatic leadership.

Evolutionary theory at its best neatly explains why some characteristic providing some beneficial function has been selected for. It is less good at explaining why some dysfunction have not been selected against. Schizophrenia as a species-dysfunction that affects 1% of individuals may simply not have a strong enough effect for a selection against the dysfunction to occur. Here we get a glimpse of another central challenge related to interdisciplinary integration, which I will discuss in more detail in the following chapters. The challenge is that of making sure that an imported element is indeed fit for carrying out the function it is assigned in the integrated approach. In the case of schizophrenia research described above, evolutionary theory is apparently not the right tool for the job at hand. For now, I will merely point in the general direction of this issue, and note that it will play a significant part in my analyses below.

This, then, is *insight A*: We should not assume that some element will work well in a new context, simply because it worked well in its original context. Recall *insight B*: We should not assume that some element will not work well in a new context, simply because it did not work well in its original context.

Some approach may well be a failure in its original application, and prove to be fruitful in different contexts. On the other hand, some approach might have been successful in its original application, and fail completely in a different context. Evaluation must, consequently, always be carried out on a case by case basis. One should be careful not to ascribe more or less validity to some theoretical element or approach than it actually deserves.

The Neurological Turn

As mentioned above there is an evolutionary subdiscipline of neuroscience. But neuroscience itself is a very popular source for inputs into interdisciplinary integration. Think of neurolinguistics, neuropsychiatry, neuropsychology, educational neuroscience, neuroanthropology, neurocriminology, neuroeconomics, neuroepistemology, neuroesthetics, neuroethics, neurolaw, neuromanagement, neuromarketing, neurophenomenology, neurophilosophy, neuropolitics, neuropsychoanalysis, neurosociology, neurotheology, and so on.

There is no doubt that the central nervous system is centrally involved in many activities in which humans (or other animals equipped with a central nervous system) engage. This does not mean, of course, that neuroscience is relevant to any subject related to humans. Neither does it mean that neuroscientific approaches (or the elements that constitute them) are equipped with universal fitness.

The popularity of importing elements from, for instance, neuroscience makes one wonder if this is not an instance of a general tendency which Alan Chalmers has diagnosed like this:

> Many in the so-called social or human sciences subscribe to a line of argument that runs roughly as follows. "The undoubted success of physics over the last three hundred years, it is assumed, is to be attributed to the application of a special method, 'the scientific method'. Therefore, if the social and human sciences are to emulate the success of physics then that is to be achieved by first understanding and formulating this method and then applying it to the social and human sciences." (Chalmers 1999, p. xx)

Chalmers is right to some extent, though it is important to be aware that "the scientific method" is not limited to methodology applied in physics these days. Certainly, neuroscience and evolutionary biology are also popular sources for certified scientificity. Importantly, Chalmers continues:

> Two fundamental questions are raised by this line of argument, namely, "what is this scientific method that is alleged to be the key to the success of physics?" and "is it legitimate to transfer that method from physics and apply it elsewhere?". (ibid.)

The reason why the latter question is too often insufficiently addressed is partly that we lack a method for detailed evaluation of specific cases of transferral. Chalmers' first query is, at least in its most obvious interpretation, a trick question. There is no single scientific method. Rather, there is a plurality of scientific methods, and whether a specific method is good depends on context. It is dependent on the purpose for which it used, the inputs it is fed, the role it is assigned, and so forth.

Empirical studies of the "seductive allure" of neuroscience has some interesting implications relevant to transferral of methodology from high rank sciences. Apparently, non-experts are highly susceptible to being unduly impressed by neuroscientific information—even when this is irrelevant to a provided explanation (Weisberg 2008; Weisberg et al. 2008). In a related study, McCabe and Castel (2008) found that their subjects were much more likely to find scientific claims and reasoning plausible if the results were accompanied by an image of a brain as compared to other types of scientific illustration (Figs. 3.1 and 3.2).

Though interesting and neatly devised, these studies have one significant weakness for our present concerns in that they do not address to what extent experts from disciplines other than neuroscience count as non-experts when presented with neuroscientific information (or brain images). Weisberg et al. compare neuroscience experts with students of neuroscience and lay men. In the McCabe and Castel study it is merely reported that the participants are undergraduate students. We are not told their subjects of study.

In science studies, Harry Collins has targeted similar issues. Ever the poet, he has coined the phrase "distance lends enchantment" (Collins 1985, p. 17; Collins and Evans 2002, p. 246). The idea has been developed further by the Edinburgh sociologist Donald MacKenzie, who reached the conclusion that scientists are often less skeptical regarding the outputs of neighbouring disciplines as compared to outputs of their own discipline (Collins and Evans 2002, p. 287; MacKenzie 1998).

Though these considerations may come across as somewhat speculative, it does not seem unreasonable to assume that scientists themselves are to some extent impressable by results from high rank disciplines. One could imagine a continuum from experts in neuroscience through closely related sciences, with scholars focused strictly on qualitative research at the other extreme. The relation is complex, though. One cannot assume that, for example, people engaged in purely qualitative research are as likely to be allured by neuroscience as lay people, not least since actively engaging in idiographic, interpretive research may correlate with a skeptical attitude towards quantitative methods. This would fit well with MacKenzie's account of what he calls "the certainty trough", according to which people far removed from the central core of scientist tend to think

Fig. 3.1   A small brain

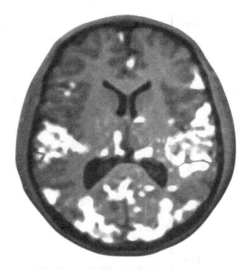

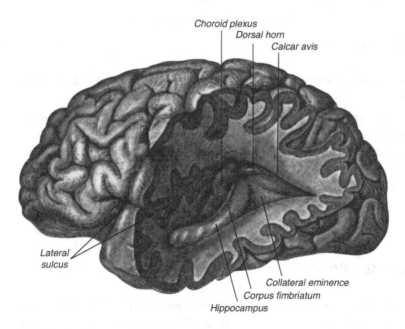

Fig. 3.2   Another small brain

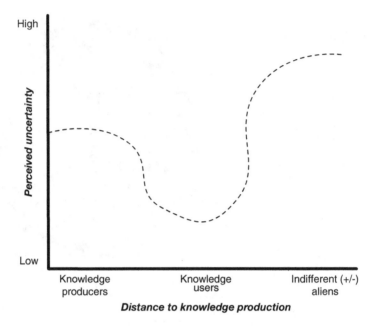

Fig. 3.3 The certainty trough

of the results as even more uncertain than the scientists themselves (MacKenzie 1998) (Fig. 3.3).

There are indirect effects to consider as well, which might draw researchers towards integration with neuroscience. For instance, scientists are likely to experience that people take their work more seriously when they add brain images to their publications and a "neuro"-prefix to their disciplinary label. Due to the pleasant nature of being admired by peers and the general public, such effects may, perhaps subconsciously, motivate scholars and scientists to engage in opportunistic neuro-interdisciplinarity.

What is the Point?

Presenting the above examples is primarily meant to highlight some of the more dominant trends in interdisciplinarity—especially in recent decades. All of the presented turns, including the turn-turn, are symptoms of a changed general approach to driving science and innovation forward. The construction of new disciplines informed by, for example, cognitive, evolutionary, and neurological approaches is a phenomenon that has co-developed with the academic interest in interdisciplinarity since Jantsch.

It would be unfair, though, if I were to leave the impression that subliminal cognitive biases were the primary motivations for interdisciplinarity. This is certainly not the case. In the next section, other reasons for interdisciplinarity will be discussed, none of which portray interdisciplinary scientists as seduced by anything at all.

More Reasonable Reasons for Interdisciplinarity

A number of non-epistemic advantages are commonly associated with interdisciplinarity. One influential view sees interdisciplinarity as raising the accountability of science by developing "novel forms of quality control which undermine disciplinary forms of evaluation" (Barry et al. 2008; Strathern 2004). To "undermine disciplinary forms of evaluation" constitutes a direct attack on the exclusive authority of established disciplines to determine intra-disciplinary matters. Sometimes this is considered a democratisation of science.

In another perspective, interdisciplinarity is viewed as a way to make science more readily applicable for innovation, which paves the way for further integrating science into the knowledge society (Nowotny 2005; Strathern 2006).

> [...] [A]t its best, interdisciplinarity represents an innovation in knowledge production—making knowledge more relevant, balancing incommensurable claims and perspectives, and raising questions concerning the nature and viability of expertise. (Frodeman et al. 2010, p. xxix)

This could be considered as a way of subordinating science to the pragmatic needs of society.

At this point, it will be helpful to look at what the interdisciplinarity movement reacts against. The opposite of interdisciplinarity is often considered to be specialisation. And indeed, in many cases, interdisciplinarity is thought to be a remedy against the negative consequences of specialisation understood as the continuing process of dividing science up into ever narrower and more isolated (sub-)disciplines.

According to Peter Weingart, science was much more interdisciplinary prior to the establishment of contemporary disciplinary divisions.

> The increasingly esoteric nature of knowledge production [throughout the eighteenth century] led to a growing distance from practical concerns and increased resistance to commercial and practical applications that had previously legitimized the utility of the sciences. (Weingart 2010, p. 6)

In Interdisciplinarity Studies, quite a lot of effort has been put into the critique of established disciplines and the related methods and assumptions. One example is when Frodeman states that endorsements of "[…] [s]pecialization and expertise are built upon two assumptions—that it is possible to get down to the bottom of things, and that it is possible to study parts of the world in isolation from the world at large" (Frodeman et al. 2010, p. xxxiv).

Frodeman does not explicate what he means by getting "down to the bottom of things" and therefore it is uncertain whether this is a necessary part of the motivation for specialised science. On the other hand, I claim that it is indeed in many cases possible to study parts of the world in isolation from other more or less related aspects. The question is rather to what extent this results in more or less "good" science.

From a pragmatic point of view, it certainly appears that many instances of specialisation have been, and continue to be, beneficial to the world at large. Think for instance of highly specialised fields such as condensed matter physics, with its many influential applications in medicine and user technology.[15] Quite a large number of the findings on which developments in these areas are based have been achieved only through attempts at isolating causal connections in specialised settings in order to develop manipulative abilities. All in all, it does not seem right to claim that there has been a decrease in, for instance, medical and technological developments based on applicable scientific discoveries since the eighteenth century. Clearly, then, specialisation has not had a completely devastating effect on the relevance of science to the general public.

On the other hand, of course, it is difficult to say whether we would have been even better off had science not been divided into specialised sub-disciplines. And there is no doubt that there is a trade-off between simplicity and cognitive economy on the one hand, and representational fidelity and rich contextualisation on the other.

The bottom line is, however, that: it is far from evident that we would be better off epistemically if we gave up on specialisation and isolation. There certainly are no legitimate reasons for a general warning against specialisation.

Specialisation vs. Integration

Given the reluctant attitude towards specialisation, one obvious reason for interdisciplinarity might be the aspiration to restore the scientific wholeness we have allegedly lost. Unfortunately, there might be serious obstacles to achieving this ideal, at least on an individual basis. There are ample reasons to doubt that it is possible today to develop a level of expertise that allows one to make genuine scientific contributions in several disciplines. Not to deduct in any way from past geniuses such as Galileo, da Vinci, or Leibniz, but the contemporary scientific world is far more complex than the one which these polymaths inhabited. Regardless of one's cognitive capacities one is certainly excused for not acquiring scientific omnicompetence today.

A less ambitious aspiration could be the development of what Donald T. Campbell has termed "the fish-scale model of omniscience". Campbell is highly critical of individual attempts to develop expertise in two or more disciplines. He calls this the "leonardesque aspiration". Campbell takes as his starting point that the specialisation of the disciplines some-how leaves gaps which are not addressed. By means of a series of diagrams he argues that when a number of closely related disciplines are organised in departmental structures, the most central of these (i.e. the disciplines that have most over-laps with the largest number of the other disciplines thus subsumed) will inevitably end up dominating the department. This will in turn hurt the connections with neighbouring departments, since the less central disciplines are those most likely to share common ground with non-central members of other departments. The result, according to Campbell, is gaps between departments in which topics are not addressed unless interdepartmental cooperation is carried out (Campbell 1969).

Campbell's own remedy for dealing with interdepartmental gaps is the fish-scale model. On this model, academic institutions should be organised in a wealth of specialised units in such a way that all units just slightly overlap with neighbouring units in a fish-scale pattern. Presumably, this should prevent gaps from developing

One might wonder, though, how actually to interpret Campbell's notions of gaps and overlaps. First, it seems that scientific activities are far too complex phenomena to be represented simply in terms of two-dimensional diagrams. Talk of gaps is obviously metaphorical, but what are the real phenomena this metaphor targets? And even if we find these real gaps, how likely is it that gaps are simply left unaddressed for long? It seems like it is often considered a central achievement for scientists to define or uncover a new problem area to discuss or analyse, to expand by applying existing theory to issues hitherto left untouched, or, indeed, to develop brand new theories or models to address issues somehow related to one's central area of expertise. One might argue that whenever a stone is discovered to be unturned, lots of ambitious scholars and scientists can be expected to be on the brink of throwing themselves at it (to turn it over) (Bechtel 1986, p. 27).

A better reason for interdisciplinarity could be to directly attack what Campbell calls the ethnocentrism of disciplines. Campbell defines ethnocentrism as "the symptoms of tribalism or nationalism or ingroup partisanship in the internal and external relations of university departments, national scientific organizations, and academic disciplines" (Campbell 1969, p. 328). Campbell suggests that fighting ethnocentrism of disciplines is a significant step towards the development of fish-scale coordination. As Gold and Gold states:

> The training and socialisation that the student in a discipline undergoes lead to an identification with the disciplinary community that comes prior to, and is generally regarded as more persistent than, identification with a specific employer or particular task. In combination with normative

differences these feelings of identification and loyalty to the disciplinary community can, and often do, develop to the point of professional chauvinism [...] (Gold and Gold 1983, p. 94)

There is certainly some truth to such claims, and consequently good reasons to consider to which extent mechanisms such as those for peer review, getting tenure or promotion, deciding on curricula, or general requirements for obtaining a degree, actually serve epistemic goals. If one wants to get a job at some point in a certain academic discipline, there is certainly a canon one is required to acquaint oneself with. Through the study of these central topics of the standard literature of the discipline, students are likely to develop some form of loyalty and trust towards other (senior) members of the discipline.[16] Special skills and values are internalised as part of ongoing activities, skills which might not be valued in other departments. Competition for funding and other resources (such as offices with a view) may also at times add to a somewhat hostile interdepartmental atmosphere.

If, on the other hand, we look in some detail at a small philosophical department to which I used to be affiliated, a quite different picture emerges. In this philosophy department, the people employed cover a vast range of interests aside from the strictly philosophical topics of their central interests. In the department, employees are quite accomplished in such divergent fields as musicology, literature theory, religious studies, psychology, psychopathology, mathematics, law studies, evolutionary theory, educational studies and more. One might wonder whether this is not the general tendency one would encounter if one were to check the extra-disciplinary interests and competencies of the scientific staff in other departments? This might constitute a fish-scale model of expertise on an individual, rather than on a departmental level. This does not mean, of course, that all employees at the department suffer from leonardesque aspirations. But it means that the department overlaps to a significant extent with many other areas of study, and even though there are central topics which are held to have primary importance by all (or most) departmental members, there is a considerable heterogeneity in secondary interests.

Three Modes of Integration

Campbell's notion of the 'leonardesque aspirations' suggests expanding a bit further on what one might call different *modes of integration*.

Let us briefly discuss three different modes. The following is by no means an exhaustive discussion of these issues. But having these distinct modes sketched out will prove useful for discussions below.

The Polymath Mode

When addressing the issue in terms of modes, we might rephrase Campbell's 'leonardesque aspiration' as 'the polymath mode'. This would be the mode in which interdisciplinary integration is carried out within the cognitive system of a single individual, so to speak. One might think of this as the mode in which one person fully specialised in a certain discipline acquires expertise in one (or more) additional discipline(s) in order to integrate the theory or methodology from these approaches.

While one might be skeptical about developing full expertise (whatever that might mean) in two or more disciplines (let alone in *all* disciplines), it is certainly possible for an individual to develop expertise *to some extent* in several disciplines. It is clear, though, that this is a very demanding approach. As everybody involved in research is aware, keeping up to date with state of the art in just a single discipline requires significant effort. Thus, keeping up to date with two or more not closely related disciplines would be a tremendously demanding task indeed. The question is, then, whether the extent to which the poly-expertise developed is sufficient for handling potential problems with sufficient care.

The Social Mode

The most obvious alternative to the polymath mode would be a social mode, in which people specialised in different discipline collaborate to combine their competencies. Representatives of different disciplines each contribute their special expertise to the common development of hybrid solutions.

This is probably the way in which most interdisciplinary research is carried out. This mode has its own challenges. Somebody will have to take upon themselves the role of mediators between the involved frameworks, and these persons will need some sort of *interactional expertise* in Harry Collins' terms:

> *In such circumstances the party without the interactional expertise in respect of the other party should be represented by someone with enough interactional expertise to make sure the combination is done with integrity.* (Collins and Evans 2002, p. 256; italics in original)

This more or less means that someone involved will have to be sufficiently "polymath-like" to carry out the function of mediator between the involved disciplines. Possessing interactional expertise must require some level of insight into the interacting disciplines. But the level of disciplinary expertise required may not be the same as that required for being a full expert in the individual disciplines.

The Educational Mode

In (socially) successful cases of interdisciplinarity, it will at some point seem natural to establish institutions at which future practitioners of the interdiscipline are educated. The educational mode makes it possible to cherry-pick the knowledge elements and competencies which people engaged in the relevant activities are required to have. It also makes possible the training of students specifically to develop the required types of interactional expertise.

All of the Above

In some cases, the above modes may be combined. For instance, a researcher who strives to become a certain kind of interdisciplinary polymath may establish a centre for the study of a given subject in his or hers preferred perspective. At this centre, researchers from the involved disciplines may be employed or affiliated on the leader's initiative. If a sufficient degree of success is achieved, this might eventually lead to the establishment of institutionalised education.

Science without a Core Set?

One central worry relates to the notion of a scientific core. Cores can be analysed along different dimensions based on what aspects of disciplinarity one favours.

Lakatos famously insisted that some theoretical and methodological hard core must be retained through changes, for some class of activities continuously to belong to a given discipline (or, rather, research programme) (Lakatos 1978). Harry Collins, on the other hand, operates with a concept of "the core set" (a more sociologically orientated core), which refers to the central group of scientists who are actively engaged in settling scientific disputes (Collins 1981, 1988).

If we combine these notions into a hybrid, let us call it '*the hard core set*', we can use this to point to a group of people with sufficient insight into the central theoretical elements to contribute actively to settling disputes in a given set of scientific activities. In some interdisciplinary settings, especially if revolutionary new approaches result from integrating distant parent approaches, one might expect that no single person possesses the competencies required for membership of the hard core set. Possessing interactional expertise is not sufficient to make one a hard core set member, since people with interactional expertise are not necessarily disciplinary experts.

In the detailed discussion of the introduction of operational definitions into psychopathology in Chap. 6, one of the most evident problems revealed is the lack of a hard core set. The people involved had little insight into the crucial differences between the state of the art or the conceptual foundations of the parent disciplines. In that case, then, when developing the operationalised psychopathology, a hybrid approach was created without an accompanying hybrid hard core set. The creators of the hybrid approach did not possess sufficiently overlapping expertise to be members of a hybrid hard core set. The consequence was that even though the creators were warned (by Carl Hempel) about a good deal of the most critical pitfalls related to the use of operational definitions, the warnings were not taken duly into account in the resulting construct.

Degenerating Hard Core Sets

We are not only concerned with whether our theoretical findings will be broadly accepted by our peers, but also with whether they will live on, so to speak. This is a respect in which interdisciplinary developments are in a much less comfortable position than, say, core developments within established disciplines. If you write a groundbreaking paper or book on a central topic in philosophy, you have thousands of dedicated philosophers already employed in philosophy departments all over the world who are, to some extent, eager to pick up (or on) your ideas and may incorporate them into their teaching and so on. If, on the other hand, your ground-breaking discoveries are within an interdisciplinary field with no established departments, like, say, neuro-inuitology, the potential for having your insights established as central among students by enthusiastic colleagues is much smaller.

Especially in the polymath mode, there is another significant risk: the hard-earned interdisciplinary expertise might evaporate completely with the retirement (or death) of the central figure(s). Even with a strong hard core set, there is a significant risk that the next generation of researchers will be less well equipped to partake in core activities or carry on the work. If well-formed educational settings have been established, on the other hand, it might be the case that second-generation researchers are in some ways even better equipped to continue the work than were the initiators of the specific interdisciplinary activity. It is a reasonable worry, though, that an XY-PhD might have less insight into deep X-issues than a straight-up X-PhD. This might eventually be problematic for third or fourth generation XY-PhDs when not even their supervisors possess deep X-insight.

Here we might have an interesting interdisciplinary auxiliary to Lakatos' notion of negative research programmes. There is a risk that interdisciplinary research programmes will undergo a degeneration due to a decline in the level of academic proficiency of its core members.

Another important aspect relates to peer review: If you are participating in the core aspects of a specific X-debate in a specialised X-journal it is very difficult to get away with making unsupported claims about central issues. The peer review institution is likely to block your paper from publication, since what counts as a peer will be an X-expert. However, given the nature of the peer review institution, it is much easier to get away with making less well supported claims when publishing on X (or perhaps XY) in a Y-journal. The reviewers in the Y-journal will be experts on Y-topics and thus know as little or less about X as the author knows about Y. It is tempting to pick peers from the hard core set of the specific interdisciplinary activity. But this is not a perfect solution, especially not if the hard core set itself is non-existing or degenerating.

Summing Up

The most important observation regarding the above examples of interdisciplinarity is that none of the texts in which these examples are discussed incorporate detailed considerations about epistemic issues related to how disciplines are integrated, what gaps are, what it means to overlap, or how to handle the threat of degeneration. The focus is almost exclusively on departmental structures, socialisation, affiliation, and organisation. The last-mentioned issues all fit naturally into a sociologically-informed account of interdisciplinarity. The first-mentioned questions are most naturally addressed within the philosophy of science. Unfortunately, philosophy of science is not frequently integrated into discussions within Interdisciplinarity Studies.

It is not up for discussion that scientific inquiry is partly a social phenomenon. Undeniably, science is carried out by communities of collaborating researchers. Thus, it is obvious that an exhaustive account of science must take account of its related social aspects (Kitcher 2001; Longino 2006). However, in Interdisciplinarity Studies, as in science studies in general,[17] arguably there is a marked tendency to overemphasise the importance of social aspects of science. Work carried out in these frameworks has certainly generated a wealth of interesting insights and eye-opening case studies. But focus is predominantly on how participants become "socialised" into disciplines as well as on the power struggles between members of disciplines, between disciplines, and between science and external stakeholders (Weingart 2010, p. 8 ff.).

As Giere wrote (commenting on the work of Bruno Latour): "One can understand the desire of sociologist for a theory of science and technology that is totally sociological. But the subject simply will not allow it" (Giere 1999, p. 63). In the next chapter, I will discuss some reasons for and against the position that philosophy has something relevant to offer for analyses of interdisciplinarity.

Notes

1. Among the latter I count works such as *The Oxford Handbook of Interdisciplinarity* (Frodeman et al. 2010), *Interdisciplinarity: History, Theory, and Practice* (Klein 1990) *Interdisciplinarity and Higher Education* (Kockelmans 1979). Among the former I count (in no particular order) people such as Robert Frodeman (Director of the Center for the Study of Interdisciplinarity at University of North Texas), Julie Thompson Klein (Former president of the Association for Integrative Studies), Erich Jantsch, Carolyn and Muzafer Sherif, Peter Weingart (Former director of the Center for Interdisciplinary Research), William H. Newell (professor of interdisciplinary studies at Miami University as well as executive director of the Association for Integrative Studies), and others. The reader might of course disagree that these are representative for Interdisciplinarity Studies or that it is reasonable at all to postulate the existence of this group of academics.

2. See for instance (Darden and Maull 1977; Mitchell 2002, 2003; Morgan and Morrison 1999)

3. Notice that in the following I will often use the phrase "interdisciplinary activities" instead of the, perhaps, more straightforward "interdisciplinary collaborations". This is a deliberate choice, since the integration of distinct disciplines does not necessarily involve more than one person, and hence may not involve any collaboration.

4. This is reminiscent of some recent discussions in philosophy warning against certain dangers of interdisciplinary collaboration. Most noteworthy, John Dupré has initiated a small field focused on so-called *scientific imperialism*, which is thought to involve forcing ones framework onto a domain of which one has little expertise (Clarke and Walsh 2009; Dupré 1995, 2001; Mäki 2013).

5. I assume that the applications of Newtonian physics on human social change, which Kellert refers to, are more recent than Darwin.

6. 'Moral philosophy' is not restricted to ethical matters, but is more properly understood as "the science of human nature". On the other hand, 'experimental philosophy' is another expression for 'natural philosophy', or, in present day terms, 'natural science'. Thus, Hume aimed to apply *the* "scientific method" to the study of human nature.

7. Some might protest against viewing Aristotle as a wannabe interdisciplinarian. Indeed, it would not be unreasonable, it seems, to take the exact opposite position: That Aristotle was perhaps the most influential early figure in the movement towards scientific specialisation.

8. Or at least were more sexually attractive in the specific EEA (i.e. Environment of Evolutionary Adaptation) of the species in question. Roughly, the EEA is the environment to which a species is adapted. The intuition is that for specific traits to have evolved, they must have served a special beneficial function in a certain environment. Verbal creativity and wit may be worthless in your local discotheque (where clearly defined muscle-groups and well-placed tattoos seem to make the biggest positive difference these days) and still have served their possessor well in the EEA.

9. As a general reflection—I am still not attempting to pass judgment on psychoanalysis.

10. *Insight A* will follow shortly.

11. Even though I do not think there exists a cognitive discipline as such. 'Cognitive' may be a free-floating additive—free for anybody to use.

12. For massively influential examples see (Lakoff 1987; Lakoff and Johnson 1980; Murphy 2002; Rosch 1973).

13. See definition of 'EEA' in note 8 above.

14. It is wise, though, to check quickly beneath the toilet seat before using the facilities in places such as Sydney (at least in my experience). Nevertheless, it is somewhat paradoxical that compared to snakes or spiders it is apparently far more dangerous to encounter a horse, a cow, or a kangaroo in contemporary Australia. But still these animals rarely inspire phobias.

15. Superconductivity, quantum computing, laser-technology, and magnetic resonance imaging are a few examples of technology in which condensed matter physics is applied.

16. In one department, they may celebrate important and influential heroes, which the "idiots" of some other departments might never even have heard of.

17. Interdisciplinarity studies is closely related to science studies. Indeed, one might think of Interdisciplinarity Studies as a highly specialised branch of science studies.

References

Aldama, Frederick Luis, ed. 2010. *Toward a Cognitive Theory of Narrative Acts.* Austin, TX: University of Texas Press.

Barry, Andrew, Georgina Born, and Gisa Weszkalnys. 2008. Logics of Interdisciplinarity. *Economy and Society* 37 (1): 20–49.

Bechtel, William, ed. 1986. *Integrating Scientific Disciplines.* Dordrecht: Martinus Nijhoff.

Boyd, Brian. 2009. *On the Origin of Stories.* Cambridge, MA: Harvard University Press.

Campbell, Donald T. 1969. Ethnocentrism of Disciplines and the Fish-Scale Model of Omniscience. In *Interdisciplinary Relationships in the Social Sciences*, ed. Muzafer Sherif and Carolyn W. Sherif. Chicago: Aldine Publishing Company.

Carroll, Joseph. 2004. *Literary Darwinism: Evolution, Human Nature, and Literature.* New York: Routledge.

Cartwright, John. 2007. *Evolution and Human Behaviour: Darwinian Perspectives on Human Nature.* 2nd ed. Basingstoke, UK: Palgrave Macmillan.

Chalmers, A.F. 1999. *What is This Thing Called Science?* 3rd ed. Indianapolis: Hackett Pub.

Clarke, Steve, and Adrian Walsh. 2009. Scientific Imperialism and the Proper Relations Between the Sciences. *International Studies in the Philosophy of Science* 23: 195–207.

Collins, Harry M. 1981. The Place of the Core-set in Modern Science: Social Contingency with Methodological Propriety in Science. *History of Science* 19 (1): 6–19.

———. 1985. *Changing Order: Replication and Induction in Scientific Practice.* London: The University of Chicago Press.

———. 1988. Public Experiments and Displays of Virtuosity: The Core-Set Revisited. *Social Studies of Science* 18: 725–748.

Collins, Harry M., and Robert Evans. 2002. The Third Wave of Science Studies: Studies of Expertise and Experience. *Social Studies of Science* 32 (2): 235–296.

Crane, Mary Thomas. 2000. *Shakespeare's Brain: Reading with Cognitive Theory*. Princeton, NJ: Princeton University Press.

Crane, Mary Thomas, and Alan Richardson. 1999. Literary Studies and Cognitive Science: Toward a New Interdisciplinarity. *Mosaic* 32 (2): 123–140.

Darden, Lindley, and Nancy Maull. 1977. Interfield Theories. *Philosophy of Science* 44 (1): 43–64.

Darwin, Charles. 1871. *The Descent of Man, and Selection in Relation to Sex*. 2 vols. London: J. Murray.

Dissanayake, Ellen. 2000. *Art and Intimacy: How the Arts Began*. Seattle, WA: University of Washington Press.

Dupré, John. 1995. Against Scientific Imperialism. In *The 1994 Biennial Meeting of the Philosophy of Science Association*.

———. 2001. *Human Nature and the Limits of Science*. Oxford: Oxford University Press.

Dutton, Denis. 2009. *The Art Instinct: Beauty, Pleasure, & Human Evolution*. 1st U.S. ed. New York: Bloomsbury Press.

Fisher, Ronald Aylmer. 1930. *The Genetical Theory of Natural Selection*. Oxford: The Clarendon Press.

Frodeman, Robert, Julie Thompson Klein, and Carl Mitcham. 2010. *The Oxford Handbook of Interdisciplinarity*. Oxford: Oxford University Press.

Giere, Ronald N. 1999. *Science without Laws, Science and Its Conceptual Foundations*. Chicago: University of Chicago Press.

Gold, S.E., and H.J. Gold. 1983. Some Elements of a Model to Improve Productivity of Interdisciplinary Groups. In *Managing Interdisciplinary Research*, ed. S.R. Epton, R.L. Payne, and A.W. Pearson. New York: John Wiley and Sons.

Hume, David. 1738. *A Treatise of Human Nature: Being an Attempt to Introduce the Experimental Method of Reasoning into Moral Subjects*. London: Printed for John Noon.

Jantsch, Erich. 1970. Inter- and Transdisciplinary University: A Systems Approach to Education and Innovation. *Policy Sciences* 1: 403–428.

Kellert, Stephen H. 2009. *Borrowed Knowledge: Chaos Theory and the Challange of Learning Across Desciplines*. Chicago, IL: University of Chicago Press.

Kitcher, Patricia. 1992. *Freud's Dream: A Complete Interdisciplinary Science of Mind*. Cambridge, MA; London: MIT Press.

————. 2007. Freud's Interdisciplinary Fiasco. In *The Prehistory of Cognitive Science*, ed. Andrew Brook, 230–249. Basingstoke, UK; New York: Palgrave Macmillan.

Kitcher, Philip. 2001. *Science, Truth, and Democracy*. New York: Oxford University Press.

Klein, Julie Thompson. 1990. *Interdisciplinarity: History, Theory, and Practice*. Detroit: Wayne State University.

————. 2008. Evaluation of Interdisciplinary and Transdisciplinary Research—A Literature Review. *American Journal of Preventive Medicine* 35 (2S): S116–S123. https://doi.org/10.1016/j.joi.2010.06.004.

————. 2010. A Taxonomy of Interdisciplinarity. In *The Oxford Handbook of Interdisciplinarity*, ed. Robert Frodeman, Julie Thompson Klein, and Carl Mitcham. Oxford: Oxford University Press.

Kockelmans, Joseph J., ed. 1979. *Interdisciplinarity and Higher Education*. Pennsylvania: The Pennsylvania State University Press.

Kuhn, Thomas S. 1962. *The Structure of Scientific Revolutions*. International Encyclopedia of Unified Science: Foundations of the Unity of Science V. 2, No. 2. Chicago: University of Chicago Press.

Lakatos, Imre. 1978. *The Methodology of Scientific Research Programmes: Philosophical Papers*. Edited by John Worrall and Gregory Currie. Vol. 1. Cambridge: Cambridge University Press.

Lakoff, George. 1987. *Women, Fire, and Dangerous Things: What Categories Reveal about the Mind*. Chicago; London: University of Chicago Press.

Lakoff, George, and Mark Johnson. 1980. *Metaphors We Live By*. Chicago: University of Chicago Press.

Lattuca, Lisa R. 2001. *Creating Interdisciplinarity: Interdisciplinary Research and Teaching among College and University Faculty*. Vanderbilt Issues in Higher Education. 1st ed. Nashville, TN: Vanderbilt University Press.

Longino, Helen. 2006. Philosophy of Science after the Social Turn. In *Cambridge and Vienna: Frank P Ramsey and the Vienna Circle*, ed. Maria Carla Galavotti. Dordrecht, The Netherlands: Springer.

MacKenzie, Donald. 1998. The Certainty Trough. In *Exploring Expertise: Issues and Perspectives*, ed. Robin Williams, Wendy Faulkner, and James Fleck, 325–329. Basingstoke, UK: Macmillan.

Mäki, Uskali. 2013. Scientific Imperialism: Difficulties in Definition, Identification, and Assessment. *International Studies in the Philosophy of Science* 27 (3): 325–339.

Mammola, Simone. 2014. Does the History of Medicine Begin Where the History of Philosophy Ends? An Example of Interdisciplinarity in the Early Modern Era. *History of European Ideas* 40 (4): 457–473.

McCabe, David P., and Alan D. Castel. 2008. Seeing is Believing: The Effect of Brain Images on Judgments of Scientific Reasoning. *Cognition* 107: 343–352.

Mitchell, Sandra D. 2002. Integrative Pluralism. *Biology and Philosophy* 17 (1): 55–70.

———. 2003. *Biological Complexity and Integrative Pluralism*. Cambridge: Cambridge University Press.

Morgan, Mary S., and Margaret Morrison, eds. 1999. *Models as Mediators: Perspectives on Natural and Social Science (Ideas in Context)*. Cambridge: Cambridge University Press.

Murphy, Dominic. 2005. Can Evolution Explain Insanity? *Biology and Philosophy* 20: 745–766.

Murphy, Gregory L. 2002. *The Big Book of Concepts*. Cambridge, MA; London: MIT Press.

Nowotny, Helga. 2005. The Changing Nature of Public Science. In *The Public Nature of Science under Assault: Politics, Markets, Science and the Law*, ed. Helga Nowotny, D. Pestre, B. Schmidt-Assmann, H. Schulze-Fielitz, and H.-H. Trute. Berlin: Springer.

Pinker, Steven. 1997. *How the Mind Works*. New York: Norton.

Price, John S., and Anthony Stevens. 1999. An Evolutionary Approach to Psychiatric Disorders: Group Splitting and Schizophrenia. In *The Evolution of the Psyche*, ed. D. Rosen and M. Luebbert, 196–207. Westport, CT: Greenwood Publishing Group.

Rasmussen, Nicolas. 1991. The Decline of Recapitulationism in Early Twentieth-Century Biology: Disciplinary Conflict and Consensus on the Battleground of Theory. *Journal of the History of Biology* 24 (1): 51–89.

Rosch, Eleanor H. 1973. Natural Categories. *Cognitive Psychology* 4: 328–350.

Shalinsky, William. 1989. Polydisciplinary Groups in the Human Services. *Small Group Behavior* 20 (2): 203–219.

Shedler, Jonathan. 2010. The Efficacy of Psychodynamic Psychotherapy. *American Psychological Association* 65 (2): 98–109.

Strathern, Marilyn. 2004. *Commons and Borderlands: Working Papers on Interdisciplinarity, Accountability and the Flow of Knowledge*. Wantage: Sean Kingston Publishing.

———. 2006. A Community of Critics? Thoughts on New Knowledge. *Journal of the Royal Anthropological Institute* 12 (1): 191–209.

Wampold, B.E., S.L. Budge, K.M. Laska, A.C. Del Re, T.P. Baardseth, C. Fluckiger, T. Minami, D.M. Kivlighan, and W. Gunn. 2011. Evidence-Based Treatments for Depression and Anxiety Versus Treatment-as-Usual: A Meta-analysis of Direct Comparisons. *Clinical Psychology Review* 31: 1304–1312.

Weingart, Peter. 2010. A Short History of Knowledge Formations. In *The Oxford Handbook of Interdisciplinarity*, ed. Robert Frodeman, Julie Thompson Klein, and Carl Mitcham. Oxford: Oxford University Press.

Weisberg, Deena Skolnick. 2008. Caveat Lector: The Presentation of Neuroscience Information in the Popular Media. *The Scientific Review of Mental Health Practice* 6 (1): 51–56.

Weisberg, Deena Skolnick, Frank C. Keil, Joshua Goodstein, Elizabeth Rawson, and Jeremy R. Gray. 2008. The Seductive Allure of Neuroscience Explanations. *Journal of Cognitive Neuroscience* 20 (3): 470–477.

Zunshine, Lisa, ed. 2015. *The Oxford Handbook of Cognitive Literary Studies*. New York: Oxford University Press.

4

The Relevance of Philosophy

In this chapter, I discuss the relevance of philosophical reflections to the study of interdisciplinarity. To some it may seem obvious that philosophy of science is relevant to discussions of interdisciplinary research. After all, it is the overall goal of philosophy of science to describe scientific activities and prescribe remedies for change to the better. The extent to which the efforts of philosophers have been successful can of course be disputed.

In (1980), Science Studies-icon Trevor Pinch wrote:

> In-depth studies of the development of particular pieces of scientific knowledge form the hallmark of recent work in the sociology of science. Broadly informed by a relativist approach, the authors of such case studies have attempted to explain scientific development in a fully sociological manner. That is, the *main explanatory weight is given to the social world* rather than to the natural world. (p. 77; my emphasis)

Later in the text, Pinch states his conviction that

> [the fact that this sociological] description is in close resonance with how the participants themselves viewed the situation [...] gives us some

© The Author(s) 2018
R. Hvidtfeldt, *The Structure of Interdisciplinary Science*, New Directions in the Philosophy of Science, https://doi.org/10.1007/978-3-319-90872-4_4

encouragement that *the account offered here is at least not as divorced from real scientific activity as that offered by philosophers.* (p. 84; my emphasis)

The conviction that philosophy is out of touch with scientific reality was a central part of the motivation for the Strong Programme in the Sociology of Scientific Knowledge since the very beginning. Inspiration from especially Thomas Kuhn[1] (Barnes 1977, p. 23, 1982; Kuhn 1962) developed into a quite dismissive attitude towards philosophy of science (Laudan 1984). Succinctly put, certain sociologists aimed to seize power over what should count as authoritative descriptions of science (Bloor 1976). This meant that philosophers had to step aside.

Similarly, in the genre known as "laboratory studies", people such as Bruno Latour, Steven Woolgar, and Karin Knorr-Cetina have argued that their micro-sociological approaches made plain the irrelevance of more traditional philosophical analysis focused on science as a rational enterprise generating knowledge and truth (Knorr-Cetina 1981, 1999; Latour and Woolgar 1986).[2] Philosophers of science were considered fundamentally misguided due to their propensity for a priori speculation.[3]

Essentially the same view on philosophy of science lives on in Interdisciplinarity Studies. In the *Oxford Handbook of Interdisciplinarity*, philosophy of science is specifically criticised for being insensitive to many central aspects of interdisciplinarity. Quoting historian Bruce Kuklick (2001), academic philosophy is claimed to have been characterised by "great technical acumen wedded to societal irrelevance" since the middle of the twentieth century (Frodeman et al. 2010, p. xxxi). And further it is claimed that:

> [...] the philosophy of science has been 'pure' for decades, built on the assumption that the epistemological aspects of scientific research can be separated from the social, ethical, political, economic, and religious causes and consequences of science. (p. xxxii)

One can certainly understand the impression that large parts of twentieth century philosophy of science have somehow been isolated from scientific reality. Nevertheless, it is most wise not to succumb to the temptation of universal generalisation. If all goes well, this book should demonstrate that philosophy of science does bring something of relevance to analyses of interdisciplinarity. And it is worth mentioning, that several of the central themes in the philosophy of scientific representation which this book draws upon were originally put forward in the early 1960s (e.g. Hesse 1963; Nagel 1961; Suppes 1960, 1962). Even Kuhn himself expressed regret that he had not placed some inclusive concept of "model" centrally in his framework of disciplinary matrices (Kuhn 1970, p. 184). These and many other examples demonstrate that it is not fair to portray philosophy of science since the middle of the twentieth century as "pure" (in a derogatory sense) and irrelevant.

Indeed, though some parts of philosophy have indeed focused too intensely on less relevant aspects of science, it is also fair to claim that over the last four decades most approaches to analysing interdisciplinarity have bent their subject somewhat out of shape. Abstracting from all but social aspects is no more a healthy move than abstracting from all but logical structures. This has been a standard criticism of the sociology of knowledge, but it is no less pertinent to the specialised study of interdisciplinarity.

More sensitive to potential problems in existing methodology, Julie Thompson Klein has pointed towards some difficulties involved in the production of new knowledge, when "[i]ndividual standards must be calibrated and tensions among different approaches carefully managed in balancing acts that require negotiation and compromise" (Klein 2008, p. 116). In line with the Cartwright quotation[4] in the opening of Chap. 1, Klein states that: "Evaluation [...] remains one of the least-understood aspects [of inter-disciplinarity]" (ibid.).

One obstacle to understanding the aspect of evaluation better, is that attention is focused on issues which fit naturally into a sociological framework. For instance, when presenting the results of the American Political Science Association's task force on interdisciplinarity, John Aldrich writes: "The attraction of interdisciplinarity often seems to be more than merely combining the insights or methods or data and their analysis, which is what the definition of interdisciplinarity requires" (2014, p. 15). Before moving on to discussions of more interesting elements, it seems required to reach an understanding of what integration itself (merely) is. However, the discussion of integration is philosophical and cannot be carried out solely in sociological terms.

Thus, to repeat myself, it is certainly worth emphasising the importance of social matters. But ignoring the complex epistemic issues related to interdisciplinarity is not reasonable. Philosophical analysis is not only relevant; it is required for an adequate analysis of interdisciplinarity. In the following, I will discuss some examples of very relevant philosophical approaches to interdisciplinarity. These examples, some of which will be elaborated in subsequent chapters, may serve as a base for the development of a more general framework for the epistemic evaluation of interdisciplinary efforts.

Relevant Philosophical Approaches to Interdisciplinarity

There are, indeed, some parts of philosophical literature which focus on interdisciplinarity in one way or the other. This includes interesting discussions of communication between scientists from different fields (Galison 1997; Holbrook 2013), discussions about implications from social epistemology for understanding interdisciplinary collaborations (Andersen and Wagenknecht 2013), discussions about whether philosophy is a necessary part of well-executed interdisciplinary collaborations or, indeed, whether philosophy itself is by nature interdisciplinary (Frodeman 2013; Fuller 2010; Hoffmann et al. 2013). These are definitely all interesting topics worthy of attention, even if they do not contribute much to the development of a framework suited for the epistemic evaluation of interdisciplinarity.

Kitcher's Historical Perspective

Taking a more critical standpoint, Patricia Kitcher has developed a very detailed analysis of the failure of Freud's project of developing an all-encompassing science of mind. Kitcher's eminent scholarly work certainly pins down important epistemic implications and sets a very high standard for investigations in this field. Kitcher's work on Freud is based on an extended historical case study, and thereby sets a great example for the utility of retrospective analyses of interdisciplinary research programmes.

Indeed, the strength of Kitcher's analysis depends partly on the ease with which one may retrospectively judge the parent approaches involved. A substantial number of the theories which Freud drew upon in his psychoanalysis have fallen from grace since then. Actually, some of those theories were already considered fundamentally flawed by experts working in the respective fields at the time when they were adopted by Freud as inputs into his own construct.

As briefly mentioned in Chap. 3, it seems clear that if one builds one's theoretical construct with, for instance, Lamarckian inheritance as a corner stone, it significantly deducts from one's credibility once Lamarck's ideas are more or less unanimously rejected.[5] This was indeed a central mistake in psychoanalysis. It is important not to confuse this with *Insight A* and *Insight B* discussed above. Those insights refer to the utility of tools and approaches in different contexts whereas the question regarding Lamarckian inheritance and Freud is about using something false as a basic premise in an argument.

Kitcher's concerns are closely related to what has been called "the threat of dilettantism", which is often expressed among less enthusiastic followers of interdisciplinarity:

> A [...] concern about interdisciplinarity, the danger of dilettantism, is raised by Nissani (1997, 212) and Bauer (1990, 113), among others. After all, there is a reason why serious fields of inquiry are called "disciplines"— their practice requires time, dedication, and indeed discipline. The demands of rigorous and specialized scholarship make it exceedingly difficult to engage responsibly with more than one discipline. Dabblers may easily be misled by superficial resemblances when they are not acquainted with the technical details wherein so much of the real effort lies. (Kellert 2009, p. 35; references in original)

Kellert is right to state that "[b]ad scholarship is the problem" but is less obviously correct when he adds that "interdisciplinarity itself carries no special risk" (2009, p. 35). Basing your efforts on work from unfamiliar disciplines is treacherous for a variety of reasons discussed throughout this book. You probably do not have sufficient insight into the various discussions of pros and cons when working with a given theory. You are not intimately familiar with the idealisations made. Further, there is the issue of keeping track of the progress of the parent disciplines, and updating your hybrid approaches in accordance with these new developments.

In addition to the difficulties related to acquiring sufficient mastery of a second discipline already discussed, there is a further significant risk that the already acquired mastery of one's home-approach may lead one to overestimate one's understanding of what is going on in a second discipline.

Indeed, this phenomenon of referred expertise has been targeted in science studies:

> At the very outset [...] we noted that in the 1950s scientists were often attributed with authority to speak on subjects outside their narrow area of specialization. [Science studies] has shown how dangerous it is to take this kind of referred expertise at face value, since the pronouncements of the wider scientific community are nearly always based on simplified and retrospectively constructed accounts of the scientific process. Quite simply, scientists' supposed referred expertise about fields of science distant from their own is nearly always based on mythologies about science, rather than on science itself. (Collins and Evans 2002, p. 259)

The strength of the threat of referred expertise depends to some extent on the mode in which the interdisciplinary activities are carried out. But the notions of referred expertise and disciplinary distance add nuance to Kitcher's warnings against building one's interdisciplinary castle on a foundation of ill-understood theory. These notions further emphasise the relevance of discussing the difference between transferring elements (mathematical or otherwise) between proximate approaches and more distant approaches by indicating that there are likely to be different grades of problems related to transferability.

Weisberg's Vehicle Perspective

As another especially interesting example, Michael Weisberg discusses a number of cases in which models are transferred between approaches. Weisberg has developed a framework for discussing such transferrals in great detail. Especially Weisberg's discussions of his concepts of "construal", "assignment", "scope", and "fidelity" highlight central and important aspects of the transferral of models between approaches. Weisberg's framework thus captures central parts of *the analysis of model use*. I shall discuss these aspects of Weisberg's account in more detail in Chaps. 5 and 7.

As Weisberg notes about the transferral of models between approaches:

> In some instances of modeling, construction is carried out from scratch. For example, when Volterra first constructed the mathematical structure for the Lotka–Volterra model, he had no previous biological models from which to work. However, in many cases, structures are developed from other structures, or may be completely coopted from another use. Any subsequent researchers using the Lotka–Volterra model are borrowing their mathematical structure from Volterra and from Lotka. (Weisberg 2013, p. 75)

At least in some cases involving the borrowing of the Lotka-Volterra model, theoretical structures are obviously transferred between disciplines (e.g. Goodwin 1967). A number of central issues are less obvious, however. Is there anything inherently disciplinary about mathematical structures? Is it relevant in anything but a historical sense that a given equation was first applied to a certain target? If not, are mathematical structures neutral? How do we compare the applications of a given equation to two different targets? Are such comparisons relevant at all?

Weisberg's work is of eminent quality and his framework certainly emphasises important aspects of the epistemic mechanisms of interdisciplinarity. However, there are details and dynamics of interdisciplinarity which Weisberg's as well as Kitcher's approaches fail to capture.

Weisberg and Kitcher both focus on examples in which the theoretical material which is imported is at the "surface" of the activities in question.

That is, as just discussed, Weisberg analyses the transferral of vehicles of representation, whereas Kitcher's focus is on the transferral of scientific products or outputs. Weisberg's account lacks attention to how other theoretical elements (tools, assumptions, algorithms) may be transferred during integration. And as Kitcher's explicitly states, she does not engage in the sort of general, abstract analysis, which is the specific concern of this book (Kitcher 1992, p. 4, 2007). Consequently, I believe there is room for considerable further development.

Pluralism and Representation

Interestingly, a number of philosophers in the tradition of the philosophy of scientific representation have worked on issues which could be interpreted as providing very strong arguments in favour of interdisciplinarity as a general strategy for scientific progress. In this literature, it is one of the most basic assumptions, that the world (and most of the phenomena that the world is made up from) is so complex that it cannot be captured adequately by any single model (or other kind of vehicle of representation). Consequently, a multitude of different perspectives are required each capturing some aspect of the total complexity (Giere 1999, p. 79).

This seems to provide all the reasons one needs for immediately adopting a general strategy of interdisciplinarity: If all phenomena require more than one perspective for a full analysis, then it is clear that we should combine perspectives and that interdisciplinarity is the way forward.

To some extent, the central notions of this book may be said to descend directly from the literature on scientific pluralism which is closely related to the philosophy of scientific representation. Philosophers such as Kellert (2009), Longino (2006), Cartwright (1999), and Giere (1999, 2006) have advocated ideas along the line that (often) "natural phenomena cannot be fully explained by a single theory or fully investigated using a single approach. As a consequence, multiple approaches are required for the explanation and investigation of such phenomena" (Kellert et al. 2006, p. vii).

I am in many ways sympathetic to this pluralistic stance. There are reasons for caution, though:

> A thoroughgoing disciplinary pluralism [...] suggests that sometimes the perspectives [involved in interdisciplinarity] do not fit nicely together on the same plane: they overlap or conflict or cannot both be held at the same time, and yet both are needed to understand the phenomenon. (Kellert 2009, p. 38)

Some questions naturally arise following such statements: What does it mean for specific perspectives to be incompatible? And what are the epistemic consequences of attempting to integrate incompatible perspectives. To provide a framework in which these and related questions can be answered is the most significant contribution I see myself as offering with this book.

According to the literature on scientific pluralism[6] it is not just the case that all phenomena require multi-perspectival explanations (in order to be fully explained, that is). It is another fundamental assumption that these different perspectives are (at least to some extent) incompatible, incommensurable, incongruent, or simply not overlapping. Think of the following: Why is it such a rare experience to encounter discussions of interparadigmatic research? Probably because it is a basic (and well-known) assumption of widespread interpretations of the Kuhnian framework that such integrations are not feasible. As I will discuss in detail, similar problems are built into representation-based accounts of science. What the consequences are of trying to integrate incompatible perspectives is an issue that is far from clear. Importantly, this central philosophical issue has not been addressed in the literature on interdisciplinarity in spite of its obvious relevance. It is, incidentally, a central aspect of what I intend to address in the following chapters of this book.

In many cases, different approaches do not combine neatly to form new, epistemically superior wholes. It may very well be the case, though, that incompatible perspectives constitute fruitful constraints on how the individual approaches might reasonably be used and interpreted. I will discuss the issues of representation, pluralism, and incompatible perspectives in detail in Chaps. 5 and 6.

The philosophy of scientific representation, then, provides both strong arguments in favour of interdisciplinary approaches and a list of serious difficulties which need to be taken into account. Similarly, there are many good reasons why the philosophical study of scientific pluralism should be integrated into the frameworks for studying interdisciplinarity. Not least since the problems addressed are absolutely central to whether interdisciplinarity is recommendable or not. Addressing interdisciplinarity on a pluralistic, representation-based background provides ample reason for a careful and sober attitude towards interdisciplinary science.

Though some philosophers of science, such as Weisberg, have dealt with the issue of using models or theories in other settings than their original field of development, only a few have discussed this in the context of interdisciplinarity. And though most of these discussions are relevant to the discussion of epistemic aspects of interdisciplinarity, no one has developed a framework able to capture the diversity of collaborations in the present-day eclectic and almost promiscuous attitude towards scientific crossbreeding.

Perhaps Sandra Mitchell's position called *integrative pluralism* is the position most closely related to the framework developed in this book. Mitchell's work is based on detailed, representation-based analyses of *narrow interdisciplinarity* (in the terms of Interdisciplinary Studies). More specifically, she deals with the integration of various levels of explanation internal to biology (Mitchell 2002, 2003; Mitchell et al. 1997). These are fascinating and important discussions very relevant to the themes of this book. But due to their specificity, they do not capture general issues regarding interdisciplinary activities involving, for instance, the humanities and the social sciences. I shall return to Mitchell's interesting work in my discussion of pluralisms in Chap. 6.

If we are to do these issues justice, a good place to start is to dig into how scientific representation in general is accomplished in specific approaches and, on that basis, spell out difficulties related to integrating these (perhaps incompatible) approaches. Once one realises that all approaches involve individual distortions, it appears that there are few reasons to assume that the integration of two distorted approaches adds up to something which is less distorted.

Summing Up

This chapter has been a short one. The reason is that it does not require much reflection to see that philosophy is actually relevant to discussions of interdisciplinarity, even though this has been passionately disputed.

Philosophical analyses of scientific representation and discussions within the framework of scientific pluralism are not just highly relevant. They will also constitute the foundation of the approach I develop in the following chapters.

Notes

1. Though a number of other philosophers(!), for example Wittgenstein, Feyerabend, Quine, and Hanson, also made their influence count (Brown 1984, p. 12).
2. To be fair, in the afterword of the second edition of *Laboratory Life* Latour and Woolgar do admit that some parts of philosophy may not be entirely irrelevant after all (1986, p. 279 ff.).
3. Straw man or not, among other interesting effects the topic stimulated a quite entertaining exchange of verbal blows between David Bloor and Larry Laudan (see Brown 1984).
4. "[W]e have no articulated methodologies for [evaluating] interdisciplinary work, not even anything so vague and general as the filtered-down versions of good scientific method we are taught at school" (Cartwright 1999, p. 18).
5. Even if one were to accept the quite controversial claim that epigenetics somehow rehabilitates Lamarck, this does not seem to help Freud. The nature of the acquired traits his theory presupposed as inheritable are more similar to the development of the long necks of giraffes than to gene expressions increasing the likelihood of obesity in offspring (Penny 2015).
6. Which I deal with in detail in Chap. 6.

References

Aldrich, John H. 2014. *Interdisciplinarity*. New York: Oxford University Press.

Andersen, Hanne, and Susann Wagenknecht. 2013. Epistemic Dependence in Interdisciplinary Groups. *Synthese* 190: 1881–1898. https://doi.org/10.1007/s11229-012-0172-1.

Barnes, Barry. 1977. *Interests and the Growth of Knowledge*. London: Routledge and Kegan Paul.

———. 1982. *T.S. Kuhn and Social Science: Theoretical Traditions in the Social Sciences*. London: Macmillan.

Bloor, David. 1976. *Knowledge and Social Imagery*. London: Routledge and Kegan Paul.

Brown, James Robert. 1984. In *Scientific Rationality: The Sociological Turn*, The University of Western Ontario Series in Philosophy of Science, ed. Robert E. Butts, vol. 25. Dordrecht: Springer.

Cartwright, Nancy. 1999. *The Dappled World: A Study of the Boundaries of Science*. Cambridge; New York: Cambridge University Press.

Collins, Harry M., and Robert Evans. 2002. The Third Wave of Science Studies: Studies of Expertise and Experience. *Social Studies of Science* 32 (2): 235–296.

Frodeman, Robert. 2013. Philosophy Dedisciplined. *Synthese* 190: 1917–1936. https://doi.org/10.1007/s11229-012-0181-0.

Frodeman, Robert, Julie Thompson Klein, and Carl Mitcham. 2010. *The Oxford Handbook of Interdisciplinarity*. Oxford: Oxford University Press.

Fuller, Steve. 2010. Deviant Interdisciplinarity. In *The Oxford Handbook of Interdisciplinarity*, ed. Robert Frodeman, Julie Thompson Klein, and Carl Mitcham. Oxford: Oxford University Press.

Galison, Peter. 1997. *Image and Logic: A Material Culture of Microphysics*. Chicago; London: University of Chicago Press.

Giere, Ronald N. 1999. *Science without Laws, Science and Its Conceptual Foundations*. Chicago: University of Chicago Press.

———. 2006. Perspectival Pluralism. In *Scientific Pluralism*, ed. Stephen H. Kellert, Helen Longino, and C. Kenneth Waters, 25–41. Minneapolis, MN: University of Minnesota Press.

Goodwin, Richard M. 1967. A Growth Cycle. In *Socialism, Capitalism and Economic Growth*, ed. C.H. Feinstein, 54–58. Cambridge: Cambridge University Press.

Hesse, Mary B. 1963. *Models and Analogies in Science*. Newman History and Philosophy of Science Series. London; New York: Sheed and Ward.

Hoffmann, Michael H.G., Jan C. Schmidt, and Nancy J. Nersessian. 2013. Philosophy of and as Interdisciplinarity. *Synthese* 190 (11): 1857–1864. https://doi.org/10.1007/s11229-012-0214-8.

Holbrook, J. Britt. 2013. What is Interdisciplinary Communication? Reflections on the Very Idea of Disciplinary Integration. *Synthese* 190: 1865–1879. https://doi.org/10.1007/s11229-012-0179-7.

Kellert, Stephen H. 2009. *Borrowed Knowledge: Chaos Theory and the Challange of Learning Across Desciplines*. Chicago, IL: University of Chicago Press.

Kellert, Stephen H., Helen Longino, and C. Kenneth Waters, eds. 2006. *Scientific Pluralism, Minnesota Studies in the Philosophy of Science*. Minneapolis, MN: University of Minnesota Press.

Kitcher, Patricia. 1992. *Freud's Dream: A Complete Interdisciplinary Science of Mind*. Cambridge, MA; London: MIT Press.

———. 2007. Freud's Interdisciplinary Fiasco. In *The Prehistory of Cognitive Science*, ed. Andrew Brook, 230–249. Basingstoke, UK; New York: Palgrave Macmillan.

Klein, Julie Thompson. 2008. Evaluation of Interdisciplinary and Transdisciplinary Research—A Literature Review. *American Journal of Preventive Medicine* 35 (2S): S116–S123. https://doi.org/10.1016/j.amepre.2008.05.010.

Knorr-Cetina, Karin. 1981. *The Manufacture of Knowledge: An Essay on the Constructivist and Contextual Nature of Science*. Oxford; New York: Pergamon.

———. 1999. *Epistemic Cultures: How the Sciences Make Knowledge*. Cambridge, MA: Harvard University Press.

Kuhn, Thomas S. 1962. *The Structure of Scientific Revolutions*. International Encyclopedia of Unified Science: Foundations of the Unity of Science V. 2, No. 2. Chicago: University of Chicago Press.

———. 1970. *The Structure of Scientific Revolutions*. International Encyclopedia of Unified Science. Foundations of the Unity of Science, V. 2, No. 2. 2nd ed. Chicago: University of Chicago Press.

Kuklick, Bruce. 2001. *A History of Philosophy in America, 1720–2000*. Oxford; New York: Clarendon Press; Oxford University Press.

Latour, Bruno, and Steve Woolgar. 1986. *Laboratory Life: The Construction of Scientific Facts*. Princeton, NJ: Princeton University Press.

Laudan, Larry. 1984. The Pseudo-Science of Science? In *Scientific Rationality: The Sociological Turn*, ed. James Robert Brown, 41–74. Dordrecht: Springer.

Longino, Helen. 2006. Theoretical Pluralism and the Scientific Study of Behaviour. In *Scientific Pluralism*, ed. Stephen H. Kellert, Helen Longino, and C. Kenneth Waters, 102–131. Minneapolis, MN: University of Minnesota Press.

Mitchell, Sandra D. 2002. Integrative Pluralism. *Biology and Philosophy* 17 (1): 55–70.

———. 2003. *Biological Complexity and Integrative Pluralism*. Cambridge: Cambridge University Press.

Mitchell, Sandra D., Lorraine Daston, Gerd Gigerenzer, Nevin Sesardic, and Peter Sloep. 1997. The Why's and How's of Interdisciplinarity. In *Human by Nature: Between Biology and the Social Sciences*, ed. Peter Weingart, Sandra D. Mitchell, Peter J. Richerson, and Sabine Maasen, 103–150. Mahwah, NJ: Erlbaum Press.

Nagel, Ernest. 1961. *The Structure of Science: Problems in the Logic of Scientific Explanation*. London: Routledge & Kegan Paul Ltd.

Penny, David. 2015. Epigenetics, Darwin, and Lamarck. *Genome Biology and Evolution* 7 (6): 1758–1760.

Pinch, Trevor J. 1980. Theoreticians and the Production of Experimental Anomaly: The Case of Solar Neutrinos. In *The Social Process of Scientific Investigation*, ed. Karin D. Knorr, Roger G. Krohn, and Richard Whitley. Dordrecht; Holland: D. Reidel Publishing Company.

Suppes, Patrick. 1960. A Comparison of the Meaning and Uses of Models in Mathematics and the Empirical Sciences. *Synthese* 12 (2/3): 287–301.

———. 1962. Models of Data. In *Logic, Methodology and Philosophy of Science: Proceedings of the 1960 International Congress*, Stanford, CA.

Weisberg, Michael. 2013. *Simulation and Similarity: Using Models to Understand the World, Oxford Studies in Philosophy of Science*. New York: Oxford University Press.

5

Representation

In the preceding chapters, I presented and discussed the motivation for developing an approach for detailed analyses of interdisciplinary research. Further, I outlined a conceptual framework for capturing the most relevant aspects of the complex phenomena targeted in this book. In this chapter, then, I present the philosophical foundation which adds the level of nuance necessary for detailed analyses of specific interdisciplinary cases. Succinctly put, the goal in this chapter is to arrive at a deflationary semantic account of scientific practice which is sociologically and historically informed and sufficiently pragmatic for capturing representational activities across the scientific board.

The central line of thought in this book is that a fruitful approach to analysing interdisciplinary activities can be developed by adapting tools from the philosophy of scientific representation. To determine the extent to which this suggestion is reasonable, clear specifications of the relevant version of the philosophy of scientific representation are required, as well as how it is to be adapted. In this chapter the main focus is on providing a response to the former half of this challenge. The main presentation of the suggested adaptations will constitute the core content of Chap. 7. Having stated this, though, it is probably clear that these two tasks cannot be completely separated.

© The Author(s) 2018   **111**
R. Hvidtfeldt, *The Structure of Interdisciplinary Science*, New Directions
in the Philosophy of Science, https://doi.org/10.1007/978-3-319-90872-4_5

To a large extent, the following account is based on Ronald Giere's view on scientific representation. In this chapter, I describe the relevant aspects of Giere's viewpoint from the early and very influential version in his *Explaining Science* (1988) up until the more fully developed versions (most notably 1999a, 2004, 2006b, 2009, 2010, 2011). Further, I describe why I consider this to be a good starting point for an analysis of interdisciplinarity as well as in which respects it is wanting. I go on to develop what I think are the necessary augmentations for the present purposes.

When discussing the background of approach-based analysis, I will largely ignore some of the most influential positions in the philosophy of science in the twentieth century. If one were to look for detailed discussions of logical empiricism, Thomas Kuhn, Karl Popper, or social constructivism in the following, one would be disappointed. The variants of the semantic view on scientific theory, which will serve as the background for the following account, were established from around 1980 and onwards (with some notable precursors in the 1960s and 1970s). The early versions of the semantic view (most notably van Fraassen 1980) were at the same time sufficiently similar in many ways to logical empiricism and sufficiently ignorant of, e.g., Kuhn's historicism for it to make sense to start my account here[1] (Giere 1988, p. 46 ff.). Relevant insights from sociological and historical approaches to the study of science only started to impact Anglo-American philosophy of scientific representation in the 1980s and onwards.

In the following, I will attempt to strike a balance between a deflationary semantic view and a pragmatic view on scientific theory. While doing so, I will not be searching for axiomatic foundations of interdisciplinarity. I will start by briefly discussing the move from the syntactical view to the semantic view on scientific theory embodied by discussions in the early 1980s. As additional background, I will discuss the move from strictly mathematical isomorphism to a socially enriched deflationary similarity-based account of modelling. Following that, I will develop a (pragmatic) use-based account of science including a specified propositional account of scientific representation.

Thus, the discussions of this chapter draw mostly on work by people associated with different contemporary or quite recent schools of thought within the philosophy of science. Among those schools are: The (so-called) semantic view on scientific theory, the (so-called) Stanford school, the (so-called) pragmatic view on scientific theory, the (so-called) model-based view on scientific theory, and some representation-based views on scientific theory.[2]

I am not claiming that a representational account exhausts all there is to say about interdisciplinarity (or science in general, obviously). But I do claim that the focus on representational activities reveals interesting, central aspects and difficulties of scientific practice in general—and will do so when applied to interdisciplinarity. Therefore, I believe that the philosophy of representation is a good place to start, when aspiring to develop means for assessment of the details and dynamics of interdisciplinary research.

That we have reached this part of the discussion means that we are about to get down to serious business. There's no two ways about it: We will now need to get our hands dirty as part of these discussions, for example, on topics such as how theories relate to their targets, whether science can reasonably be construed as unitary along a representational dimension, and what the consequences of this might be for analysing interdisciplinarity. These are, indeed, daunting tasks, so we better start right away before we lose our nerve.

The Basics

As should be evident by now, the basic focus of this book is on activities of representation in analyses of interdisciplinarity. The thought that representation in one way or the other should be placed centrally in analyses of scientific activities in general has been a basic assumption in a quite heterogeneous set of philosophical approaches to science for quite a while. The stance that vehicles of representations (in my terms) should take centre stage is characteristic of early proponents of the semantic view of scientific theory such as Patrick Suppes and Bas van Fraassen (both considered early proponents even though their major contributions are several decades apart) and more recent contributions by, for instance, Giere, Weisberg, Downes, Godfrey-Smith, and Thomson-Jones (aka

Jones). As briefly mentioned above, it is worth noting that, for instance, Ernst Nagel (1961), Mary Hesse (1963), and, indeed, Thomas Kuhn (1970) have expressed similar lines of thought, though they are less frequently discussed in the canonical literature on the topic.

The development of the semantic view is often considered a crucial step away from the syntactical view on theory associated with the logical empiricists and kindred spirits. Indeed, the semantic view is often explained in contrast to the syntactical view, as in the following quotation:

> The syntactic picture of a theory identifies it with a body of theorems, stated in one particular language chosen for the expression of that theory. This should be contrasted with the alternative of presenting a theory in the first instance by identifying a class of structures as its models. In this second, semantic, approach the language used to express the theory is neither basic nor unique; the same class of structures could well be described in radically different ways, each with its own limitations. The models occupy centre stage. (van Fraassen 1980, p. 44)

For the purpose of analysing and comparing representational activities in a very broad-scoped construal of science, the view of modelling in early versions of the semantic view is far from adequate. Early proponents of the semantic view implicitly demarcated science as activities with a weighty mathematical core. This partly motivated van Fraassen to state that "[...] the usages of models in metamathematics and in the sciences are not as far apart as has sometimes been said" (1980, p. 44). Suppes made the even stronger assertion "[...] that the meaning of the concept of model is the same in mathematics and the empirical sciences" (Suppes 1960, p. 289). As has been frequently pointed out in the more recent literature, this is not a reasonable view of science in general. Many activities, which one might be inclined to label as "scientific", involve no mathematics at all. Stephen Downes is one influential example of philosophers disagreeing with van Fraassen and Suppes on this issue:

> There are many referents for the term "model" and it is my contention that there are far greater differences between models in mathematics and logic and models in science than holders of the semantic view have been

prepared to admit. Both Suppes and Van Fraassen have played down the distinction between models in logic and in science. (Downes 1992, p. 144)

Indeed, the statements of van Fraassen and Suppes are quite peculiar given that both of these authors discussed a row of different model types in the paragraphs leading up to the above quotations. The models they each discussed include physical models (i.e. a model aircraft), rational choice models in game theory, and the Bohr model of the atom, all of which seem easily distinguished from what is denoted by the term 'model' in metamathematics. In the discussions of modelling dominating contemporary philosophy, it is widely assumed that all these types of models (and others as well) must be encompassed by any adequate account of scientific representation (e.g. Godfrey-Smith 2009; Weisberg 2013).

As developed in his *The Scientific Image* (1980), van Fraassen's Constructive Empiricism is constructivist in the sense that scientific models are viewed as human constructs incorporating postulated elements and structures which cannot be deduced from observation. The account is empiricist in the sense that models are to be evaluated strictly on the adequacy between their empirical substructures and observable phenomena. What counts as observable is a controversial issue, which I shall not grapple with here.[3] The point is, that on van Fraassen's account, in validating a model one need not be concerned with what Giere calls the theoretical superstructures of models, that is, the within the theory postulated non-observable relations between, and causes of, observable phenomena (Giere 1988, p. 49).

On van Fraassen's account in (1980), then, a model is accepted if it is empirically adequate, which means that its empirical substructures are isomorphic with the observable data, which van Fraassen also terms 'appearances' and defines as "The structures which can be described in experimental and measurement reports […]" (1980, p. 68 f.). Empirical adequacy is far from being a weak demand for theories, but it is a logically weaker demand than the realist's demand for truth, which would implicate that every detail of a model should replicate the relevant aspects of the world—including the theoretical superstructures (at least according to van Fraassen). The distinction between those two demands is equivalent to Michael Weisberg's (2007) distinction between "representational fidelity" (corresponding to empirical adequacy) and "dynamical fidelity" (corresponding to the realist position). Weisberg's position will be further discussed below in this chapter.

The Scientific Image did not constitute a first or decisive strike against the dominance of Logical Empiricism. Quite a number of very influential discussions in the preceding twenty years had significantly reduced the authority of this attitude to philosophy of science, though on different grounds. The historicist and sociologically informed reasons to doubt the modus operandi of logical empiricists, however, had little impact on early versions of the semantic view, as already stated. Perhaps this is part of the reasons why the contrast between the semantic and syntactic "pictures" of science is less sharp than one might initially think. In the words of Gabriele Contessa:

> Philosophers of science are increasingly realizing that the differences between the syntactic and the semantic view are less significant than semanticists would have it and that, ultimately, neither is a suitable framework within which to think about scientific theories and models. The crucial divide in philosophy of science, I think, is not the one between advocates of the syntactic view and advocates of the semantic view, but the one between those who think that philosophy of science needs a formal framework or other and those who think otherwise. (2006, p. 376)

Of the two groups defined by Contessa, I belong to the group of people who think otherwise. Of course, one must acknowledge the importance of the very influential contributions by Suppes and van Fraassen. Nevertheless, for present purposes the most significant contribution to philosophical representation-based thinking is *Explaining Science* (1988) by Ronald N. Giere.

Enter Ronald Giere

In 1988, Ronald Giere published *Explaining Science—A Cognitive Approach*. Giere's account of science in this work is situated firmly within (or perhaps built upon) the semantic view on scientific theories by the fundamental assumption that *models* as representational entities are the central aspects of scientific activities. Further, Giere's focus on mathematical models, data models, and experimental models relates him closely to his predecessors within the semantic view. So do central questions such as "how do theories explain and confirm data (and vice versa)?" and "How do theories shape and restrain data (and vice versa)?".

Nevertheless, Giere's early version of the semantic view made significant departures from the track laid out by van Fraassen (and others). In (1988) Giere attempted to develop a sociologically and cognitively informed philosophy of science, drawing inspiration and challenges from SSK, STS, and, for example, the extended mind hypothesis. Giere's "bridge-building" aspirations extend into his more recent work as well (2006b, p. 3). In its more recent developments, Giere's construal of science has gained quite a significant pragmatic touch with a strong focus on how models are *used*.[4]

Constructive Realism

In Giere's view, the somewhat implausible claim of close resemblance between the meanings of 'model' in metamathematics and in the sciences was weakened to some extent. Rather than claiming that scientists' use of 'model' is *the same* as in metamathematics, he stated that it "overlaps nicely with the usage of ['model' by] logicians" (Giere 1988, p. 79).

Giere approved of van Fraassen's constructive elements but disliked the accompanying anti-realism. That is why he at first dubbed his alternative version 'constructive realism' only later to change it into 'perspectival realism'. What is interesting in this contest, however, is not the realism/anti-realism debate and the disagreements between Giere and van Fraassen along this dimension. What is most relevant is their differences regarding the notion of "model" and how to think about the relation between models and targets.

The central contributions of *Explaining Science* is, first, the idea that we should think of models "as *abstract entities* having all and only the properties ascribed to them" (Giere 1988, p. 78, italics in the original) in the scientific texts. Scientific texts, according to Giere's account then, contain statements that describe models. Secondly, according to Giere's account, there is a relation of definition between statements and model, and a relation of similarity between model and the targeted real world system. As Giere pointed out, however, in many cases models do not seem to bear close relations to any empirical systems and, consequently, it may require considerable effort to establish similarity relations (Downes 1992, p. 146; Giere 1988, 2008, p. 126).

There are substantial difficulties with Giere's notion of "abstract entities". It is quite difficult to figure out what ontological status these entities are supposed to have. As Thomson-Jones has pointed out, on the one hand it seems like Giere understands abstract in the sense of "existing outside space and time". However, it is not clear how something can have properties such as "being extended" and "periodicity" when existing outside space and time (Thomson-Jones 2010). Even though there might be solutions for this problem, they come at a considerable price. Paul Teller has suggested considering the abstract entities as objects containing uninstantiated properties (2001). But that solution simply pushes the burden of controversy on to issues regarding the ontological status of properties themselves. Peter Godfrey-Smith's alternative suggestion is to think of models as "imagined concrete things" rather than abstract entities. As Godfrey-Smith acknowledges, however, this solution is still saturated with metaphysical puzzles (some of which are identical to those haunting Giere's abstract entities) (Godfrey-Smith 2006, p. 734 f.).

For present purposes, however, we need not quarrel with these issues. The ontological status of models is not crucial to the present analysis, and we may safely proceed to exploit the overall structure of Giere's account of scientific representation without worrying about ontological details. Since the central interest presently is to track transformations or changes between parent approaches and integrated approaches, we do not need to pass ontological judgments. I believe Godfrey-Smith is right when he claims that "the treatment of model systems as comprising imagined concrete things is the "folk ontology" of at least many scientific modelers" (Godfrey-Smith 2006, p. 735).[5] On the other hand, discussions of the ontological status of models is probably rarely a part of discussions anywhere outside of philosophy of science. Indeed, I believe there is reason to doubt whether one's position on these issues makes much of a difference to (scientific) practice anyway.

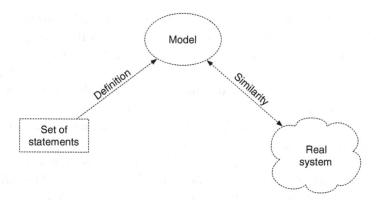

Fig. 5.1 Similarity and definition

Figure 5.1 illustrates Giere's vision of the relation between statements, models, and reality (Giere 1988, p. 83).

Extremely important for the utility of Giere's approach in the present context is that the defining set of statements can be statements expressed in any language. Some models are certainly defined (in part) in the language of mathematics, other models are simply expressed in a natural language such as English. Giere does not discuss non-mathematical representation in the social sciences or the humanities, but there is nothing to prevent us from expanding the scope of the sets of statements to include statements one might find in those circles.

Giere developed the notions of "interpretation" and "identification" as more useful alternatives to the correspondence rules of Logical Empiricism (Giere 1988, p. 75). "Interpretation" in Giere's universe means linking mathematical symbols to concepts, while "identification" refers to the act of linking these concepts to specific objects. Activities of interpretation and identification are important places to look for changes as the consequence of interdisciplinarity, though we need to include non-mathematical symbols on the modelling side of these relations.

As I understand Giere, the notion of interpretation is somehow part of the act of definition, which serves to point out the abstract entity (that is: the model) in question. Establishing the similarity relation between abstract model and real system, then, presupposes identification of the concrete aspects of the real world which the claims of similarity involve.

The move from construing the relation between model and target in terms of the softer notion of "similarity", rather than the stronger "isomorphism" (or other strong alternatives), has been very contested. Many find that "similarity" is unfit for the function it is assigned by Giere due to its vagueness. Others contest "similarity" on grounds of it implying a symmetrical relation (if the model is similar to the target, then the target is equally similar to the model), and this kind of symmetry is considered inappropriate in a scientific representational relation.

I will discuss these issues in more detail below. But let me note for now that I am quite sympathetic to Giere's use of "similarity". "Similarity" seems able to capture a very broad range of relations from different scientific practices. As Downes argues, the concept of "isomorphism" is actually relaxed considerably when used to describe the relation between scientific models and empirical systems compared to its use in metamathematics (1992, p. 145). One might argue that these different versions of "isomorphism" are actually simply more or less strict subtypes of similarity, with the metamathematical version being the strictest.

Claims of similarity might often require explication, but that is, I believe, part of a healthy process of identification (in Giere's terms) or construal (in the terms preferred by Weisberg (2013) and Godfrey-Smith (2006)). Spelling out in which ways one considers a vehicle of representation to be similar to its target seems to be a central part of sound representational practice. And there is little doubt that similarity relations are candidates for being transformed as part of an interdisciplinary process.

Perspectival Realism

An important part of the present analysis of interdisciplinarity based on representation is Giere's concept of "perspective". This concept has played a significant role in Giere's most recent account of science, which he has named *Perspectival Realism* (not to confuse his later more developed position with his earlier *Constructive Realism*) (Giere 1999a, 2006b). I will postpone the detailed discussion of the notion of "perspective" and its consequences for interdisciplinarity to the next chapter. Right now, I will focus on another aspect of the discussion of representation in Giere's later development of his account of science, namely his discussion of the need for a four-place representational relation.

Recall van Fraassen's *Hauptsatz* quoted in Chap. 1:

> There is no representation except in the sense that some things are used, made, or taken, to represent some things as thus or so. (2008, p. 23)

If we understand representation as an action, representation must always involve an agent: *Someone* uses *A* to represent *B*.

In the later versions of his account, Ronald Giere operates with a very similar agent- and action-based conception of scientific representation—though he stresses the need for a fourth element: purposes (Giere 2004, 2010). Giere suggests the following four-place-relation (Fig. 5.2) as containing the variables minimally required for an analysis of a scientific representational relation.

This schema is to be understood in the following way: A scientist, or a group of scientists (*S*), uses something (*X*) to represent an aspect of reality (*W*) for one or more specific purposes (*P*) (Giere 1999b, 2004). Notice the close fit between Giere's representational relation and Bechtel's and Toulmin's characterisations of disciplines as discussed in Chap. 2.

The four-place relation is very central to Perspectival Realism: It shows that scientific representational activities are always to be understood relative to the involved subject(s), object(s), and purpose(s). On the one hand, as I will discuss in detail in Chap. 6, representation always involve a view from somewhere. On the other hand, representational activities are always directed against an object, which, if it does not exactly determine the character of the activities of representation, at least often (and ideally) strongly affects it. Moreover, the purpose for constructing models in the first place has considerable influence on which perspectives are chosen and which results are accepted.

> S uses X to represent W for the purpose P

Fig. 5.2 Giere's representational relation

It is further important to be clear about what might take the place of the variable X. Giere's position is this:

> So here, finally, we have a candidate for the X in the general scheme for representation [...]: Scientists use models to represent aspects of the world for various purposes. On this view, it is models that are the primary (though by no means the only) representational tools in the sciences. (Giere 2004, p. 747)

So, X is a placeholder for models (among other things[6]). Giere's later concept of "model" is not identical with the *abstract entities* of his younger days. For instance, in his (2006b) incarnation, Giere considered models within the sciences to be forming a "quite heterogeneous class including physical models, scale models, analogue models, and mathematical models, to name a few" (p. 62 f.). "Models as abstract entities" still have a role to play as a subset of the generic class of models (Giere 2008, p. 125). This is a move in the right direction. But there is still some way to go before we have an operational notion of "vehicle of representation" which will serve our present purposes. I will discuss in some detail just below how Giere's later notion of "model" can be developed into a fully liberated and enriched X-variable, which is simply indicating the use of "something" as vehicle of representation.

But first, what about W? Like many a philosopher of science, Giere is mainly occupied with understanding the natural sciences. But it is worth emphasising once again that nothing prevents us from expanding our understanding of W to include all kinds of phenomena, such as the friendship of Watson and Holmes, the influence of meteorological circumstances on the choices of colour among impressionistic painters, as well as the movement of a pendulum or the function of a Higgs-field. It is important not to let the focus on natural science and representation of "real systems" mislead us into thinking that we are only addressing (inter-disciplinary) studies of "real" physical stuff.

As Downes has stated: "There are many models in science that clearly do not purport to represent empirical systems and yet are still important in scientific theorizing" (Downes 1992, p. 143). Similarly, Thomson-Jones has an interesting discussion of the representation of "missing systems"

(2010). To develop a framework for capturing general dynamics of inter-disciplinarity, it is vital not to exclude potential participants in interdisci-plinary activities. On the other hand, there is little reason to fear being overly inclusive. If people want to study the emotional states of quarks or the childhood traumas of Batman, who can decide, a priori, that these studies are unscientific or cannot be studied better (or worse) in some interdisciplinary setting?

In the next section, I will suggest moving focus from *models* and towards *modelling*. In this process, the concept "model" will be developed further. "Model" is not the perfect concept to place at centre stage in the discussion of interdisciplinarity. But the following discussion of model-ling throws valuable light on how best to construe interdisciplinarity qua representational activity.

The Expanded and Enriched *X*

Part of the discussion within the semantic view has been focused on try-ing to capture the nature of theories. Indeed, Giere has invested some energy in discussing how populations of some cluster (of clusters of clus-ters) of models translate into some specific theory.

> My preferred suggestion, then, is that we understand a theory as compris-ing two elements: (1) a population of models, and (2) various hypotheses linking those models with systems in the real world. Thus, what one finds in textbooks is not literally the theory itself, but statements defining the models that are part of that theory. One also finds formulations of some of the hypotheses that are also part of the theory. (Giere 1988, p. 85)

Luckily, for present purposes we need not concern ourselves further with issues about the nature of theories. Unfortunately, though, we need to be a lot more specific about the links between vehicles of representation and targets, because elements of these links may be part of what is mixed up during an interdisciplinary process. Focusing on these links will help us determine important aspects of the pre- and post-interaction state of affairs of an interdisciplinary integration. The next step, of course, is to determine whether or not the integration brings us in a better or worse epistemic position.

In the following sections, I will start out by briefly discussing Stephen Downes' (1992) deflationary account of modelling and theorising. Next, I will discuss Martin Thomson-Jones' suggestion for a propositional view of modelling. If we are to compare vehicles of representation used in all types of science potentially involved in interdisciplinarity, we need a category of vehicles of representation with a very broad scope. Downes and Thomson-Jones both add valuable elements to the development of an adequate account of modelling.[7]

Deflation

Stephen Downes' account of theorising as presented in (1992) takes the crucial step away from attempting to define what theories (or models) *are* to a focus on the activities of theorising and modelling. As he states: "[...] we cannot say what scientific theories are unless we appreciate the myriad ways they are used and developed in all of the sciences" (p. 142). As Downes mentions briefly, in spite of extensive philosophical discussions of the nature of scientific theory over the years, little effort has gone into establishing agreement on the scope of theories or models. What are the minimum requirements for something to be a scientific theory or model? It seems like any boundary drawn is either obviously wrong (like on the received view) or simply arbitrary.

Currently, I have no ambitions of contributing to discussions of demarcation. For present purposes, the best we can do is to bracket questions about the nature of theories and models and simply count any act of representation in any alleged scientific context as worthy of inclusion in our analysis of interdisciplinarity. Further, we should pay no regard to the nature of whatever is used as vehicle of representation in the context. The only requirement must be that it is possible from the descriptions provided by the participants in the scientific activity in question to figure out what is indeed used as vehicle of representation, and what it is used to represent. In many cases, one might end up concluding that how some vehicle is supposed to represent some phenomena is radically underdetermined, and that the act of representation is therefore impossible to evaluate. But this will in itself provide some hint of how to assess the scientific activity in question, I believe.

In the literature on model-based accounts of science since 1980, it is quite clear that there is a tendency towards expanding the category of models. As Downes points out, Giere contributes considerably to a proliferation of various types of models, for instance when he discusses the modelling involved in the geological revolution (Giere 1988, p. 227 ff.). Griesemer adds physical models (such as Watson and Crick's cardboard and wire model of DNA) as well as what he calls "remnant models" (such as museum specimens of animals or plants) (1990). Michael Weisberg discusses examples such as the San Francisco Bay Model (the warehouse-size model simulating the consequences of building a dam in the San Francisco Bay), The Moniac (the analogue simulation of the british economy by means of "fluidic logic"), as well as model organisms (such as lab rats) among others (2013).

Downes ends up concluding very minimally that modelling, in one way or the other, is an important part of scientific activities. I think this deflationary view is reasonable. We might still, though, benefit from using Giere's schema of representation when analysing what is going on in cases of interdisciplinarity as I shall demonstrate below.

Downes expresses his opinion in the following way:

> Although Giere claims that "theories are families of models", his view could be more adequately characterized as the deflationary: model construction is an important component of scientific theorizing. (Downes 1992, p. 150)

I will twist this a bit and state that *representation is an important part of scientific activity* and, further, that *determining how vehicles are used is an important part of the analysis of representation*. For an account of vehicles of representation and their use to be rich enough for present purposes, it is required that we look beyond the vehicles themselves to the processes of identification and interpretation (or construal). We must look in more detail especially at how X's are linked to W's. Martin Thomson-Jones' propositional view of modelling (2010, 2012) points in the right direction. He does not quite make it home, though, so we will have to move further ahead, once we have gone through his position in the next section.

To Model (Mathematically) or Not to Model (At All)

As already discussed at length, 'model' is the standard term for denoting "the things by which other things are represented" and also by far the most discussed type of vehicle of representation (Cartwright 1983; Giere 1999a, 2006b; van Fraassen 1980, 2008; Weisberg 2007). I believe that the generic expression 'vehicle of representation' is better suited for capturing representational practices, but nevertheless there is little doubt that the use of models and the practice of modelling is widespread in science. This is especially evident if one adopts a liberal understanding of 'model' (as I have argued in favour of just above).

Giere's focus on how Newton's laws translate into model-based representation of real world systems provides us with some significant insights. But zooming in on classical physical modelling partly occludes that scientific representation is carried out by means of a multitude of vehicles, only some of which are mathematical (or non-linguistic abstract entities for that matter). It is especially worth noting that many vehicles in the natural and health sciences are far from abstract, and that many vehicles in the humanities appear to be of a very linguistic nature indeed. This however does not detract from Giere's overall account. One can reasonably maintain his description of a set of statements describing (perhaps even defining) the vehicle of representation.

But there are other relevant issues to discuss in this respect. Because how is it, exactly, that the X and the W is linked on Giere's account of the representational relation? We need to get a hold on what is involved in the identification process if we are to compare different approaches in the pre- and post-interaction states of affairs. Thomson-Jones' (2012) argument in favour of a propositional view of modelling points out relevant aspects to consider.

Thomson-Jones bases his discussion on a distinction between *mathematical models* and *non-mathematical models*. By 'mathematical model' he simply means a model which incorporates some mathematical tools. To Thomson-Jones this does not imply that the model *itself* is a mathematical structure, though. So, in this way Thomson-Jones' notion of "mathematical model" is significantly weaker compared to some widespread understandings of what mathematical models are.

According to Thomson-Jones, most (perhaps all) mathematical models include sets of propositions that indicate, among other things, how the involved mathematical structures relate to their target system(s). Non-mathematical models, then, consist of sets of propositions (and perhaps diagrams or other illustrations). Thomson-Jones quotes Downes for an example of a non-mathematical model:

> [...] "[I]n most texts a schematized [eukaryotic] cell is presented that contains a nucleus, a cell membrane, mitochondria, a Golgi body, [the] endoplasmic reticulum, and so on" (1992, p. 145). This model is not a model of any particular cell, or even of any very specific type of cell. It is typically presented by means of a diagram and some surrounding text describing, for example, the functions of some of the organelles the cell contains. Crucially, no mathematics is employed. (Thomson-Jones 2012, p. 762 f.; reference to Downes in original)

Thomson-Jones briefly discusses other examples of non-mathematical models picked from biology, evolutionary theory, physics, and chemistry. It is, I take it, quite evident that if one were to look at less paradigmatic scientific disciplines, one would find an even greater abundance of cases of non-mathematical modelling (while mathematical models would be scarce).

The textbook model of cell structure involves no mathematical tools. If still we think of it as a model, its content must be of a different kind.

> Instead, I propose that the textbook model of the cell is a collection of propositions. The following propositions are among those that make up the model: that the eukaryotic cell has a membrane, that it has a nucleus, that the nucleus contains a nucleolus, that the nucleolus has such-and-such functions, and that the cell contains mitochondria. These are some, but not

all, of the propositions contained in the collection of propositions that we can take the model in question to be. (Thomson-Jones 2012, p. 764)

Thomson-Jones names his position *the propositional view*. On Thomson-Jones' account, then, it is possible to have models that involves no mathematics at all. But even when discussing this, he remains in close proximity to the most paradigmatic and respectable types of science. This is wise, I think, in the sense that he thereby indicates that the non-mathematical aspects of his account are relevant and important even in the cases where they are least likely to be warmly welcomed. He even to some extent safeguards against this kind of resistance by stating that:

> [...] we should bear in mind that if we want to understand the epistemology and methodology of modeling as an activity, then it will surely be important to understand all the stages of the modeling process, and even when the outcome of that process is a mathematical model, there will often be nonmathematical models lurking in the prehistory. This point provides us with another reason to take nonmathematical models seriously and to insist on an account of the nature of models that can accommodate them. (Thomson-Jones 2012, p. 771)

On the other hand, he fails to address to what extent his argument opens for very non-exemplary cases to be included in his category of models. As examples of other things which might be considered models on the propositional view consider, for example, models for literature analysis (e.g. the actantial model), models in psychology (e.g. the diathesis-stress model), models for stakeholder analysis (e.g. the Power/Interest-grid), or models for analysing corporate strategy (e.g. SWOT-analysis or The Strategy Diamond).

In my opinion, the inclusiveness of Thomson-Jones' account is a virtue. Indeed, the propositional view on models is especially useful in relation to a general method for analysing interdisciplinarity for which one needs to be able to encompass miscellaneous scientific approaches. Nevertheless, Thomson-Jones' account of modelling ends up being somewhat messy, I believe, as exemplified by the following quotation:

> [...] mathematical modeling (often, and perhaps always) involves constructing a collection of propositions that are, in part, about relations

between a certain mathematical structure and the target system, and we then use the term 'model' to refer both to the collection of propositions and to the mathematical structure. (Thomson-Jones 2012, p. 767)

When Thomson-Jones describes models in this way, his concept of "model" comes very close to my notion of "approach". This is, however, a potentially confusing way to use 'model'. It is much clearer to separate out vehicles of representation and the intermediate layer as I have suggested above and then use 'approach' to refer to the two combined. My use of 'vehicle of representation' rather than 'model' has the dual function of highlighting the inclusiveness I prefer, while reducing the likelihood of infuriating people who are very devoted to a specific (less inclusive) sense of 'model'.[8] Separating out vehicles and intermediate layers further makes it easy to consider cases in which, for instance, a specific vehicle is retained even though the approach is altered. Such situations are certainly less straightforward to handle if the entire mess is denoted by 'model'.

The Propositions

Which are these underlying propositions in which (perhaps all kinds of) models are embedded? This is a good question, not only since it is largely unaddressed by Thomson Jones' treatment, but also since a good answer is needed for solving some difficulties haunting existing theories of representation, such as the so-called *bridging problem* (i.e. "the problem of how to bridge the gap between models and the world" (Contessa 2010, p. 516)).

Nancy Cartwright writes of bridging principles as that which connects the deepest (most abstract) theoretical layers (such as basic laws) with empirical systems (Cartwright 1983, p. 143 f.). Cartwright's discussions are extremely apt and insightful, but they relate entirely to disciplines in which the empirical systems are neatly quantified. Consequently, the gaps that need to be bridged are to a large extent internal to mathematics. The bridging problem is relevant to all disciplines. But in some disciplines the conceptual gaps to be bridged are decidedly wider. Perhaps, for present purposes, it makes better sense to talk of *bridging procedures*, since 'principle' has a certain feel of *a priority* and *explicitness* to it. In many cases, it is far from explicit how gaps are bridged. Nevertheless, transformations of bridging procedures might be one outcome of inter-disciplinary integration, and therefore we need to be able to figure out the bridging procedures in parent approaches involved in interdisciplinarity, and to what extent these bridging procedures are transferred or transformed during integration.

Mary S. Morgan and Margaret Morrison have suggested that modelling involves "fitting together [...] bits which come from disparate sources" (Morgan and Morrison 1999, p. 15). In the same volume Marcel Boumans suggested a variety of factors involved in the construction of models, such as metaphors, analogies, policy, stylisation, as well as mathematical techniques and concepts (Boumans 1999, p. 93). The elements involved and their importance supposedly vary with disciplinary context.

My position is that the *underlying* propositional content (i.e. what I call "the intermediate layer") can be fruitfully thought of in terms of bridging procedures (constituted by, among other things, Boumans' factors). Further, I suggest that these bridging procedures can to a large extent be characterised in terms of tools, propositional algorithms, and assumptions as discussed earlier (in Chap. 1. Further details await in Chaps. 6 and 7). Spelling out the involved elements is certainly not always easy or straightforward. But it is possible in most cases—at least to some extent. Further, I argue that in cases where this is impossible, serious doubts are raised about the trustworthiness of the scientific activities in question. If it is impossible to figure out how a vehicle of representation is linked to its target, it is hard to imagine how else one should evaluate the epistemic aspects of such a relation.

Weisberg on Construal; Assignment; Fidelity

Michael Weisberg makes use of some notions which may add useful nuance to the discussion at this point (Weisberg 2007). In (1988, p. 75; see also 2006b, p. 64) Giere introduced the distinction between interpretation and identification which can be considered as the general framework behind Weisberg's "construal". The interpretation of a model is the general assignment of the kind of thing represented by some parameter of a model (e.g. "x represents that weight of a falling object"), while identification is the assignment of a specific object as that which is represented (e.g. "x represents the mass of this particular object in this particular context"). Changes in both respects are relevant to the analysis of an instance of interdisciplinarity.

Drawing on Weisberg, the following can be considered as adding detail to the process of identification. First of all, Weisberg's notion of the "construal" of a given model involves the *assignment* of which parts of the model are to be used in a given act of representation. The assignment also involves explicitly pointing out which aspects of the model are to be ignored in this particular use. Next, the specification of *scope* means pointing out which aspects of the targeted phenomena are supposed to be represented by the model. To this, one may reasonably add considerations of how "specifically" this process is carried out (that is, one may consider the level of accuracy of assignment and scope in the construal of a given model). Further, 'explication' may be used to refer to the process of making clear the construed internal relations in the model as well as the relations (bridging procedures) between the model and the target (including the involved processes of idealisation which will be discussed in Chap. 6).

Finally, Weisberg operates with two fidelity criteria which determine how tightly the model must fit its target in order to count as adequate. *Dynamical fidelity* concerns to what extent the model succeeds in predicting the behavior of the target system. *Representational fidelity*, on the other hand, concerns whether such predictive success is achieved for the right reasons, i.e. whether the structure of the model actually fits the causal structure of its target. A demand for high representational fidelity would obviously be much stronger than a demand for high dynamical fidelity. As mentioned above, ideal dynamical fidelity corresponds to van Fraassen's empirical adequacy, while ideal representational fidelity corresponds to the stronger demands of the realist. The difference is, of course, that dynamical as well as representational fidelity are scaleable criteria, and that an investigator is free to determine the level of either required in a given context.

In the same vein, Giere plausibly considers the standards by which it is determined whether some approach is included in a theoretical cluster to be socially constructed (1988, p. 86). It is reasonable, I believe, to claim that interdisciplinarity at least sometimes involves (temporarily?) lowering standards for inclusion. Some ways of lowering this threshold would be to relax the requirement for the specification of bridging procedures, the requirements regarding fidelity criteria, or the required specificity of one's construal. Lowering such thresholds (temporarily) is not necessarily detrimental to the quality or fruitfulness of scientific activities, but it is certainly not an obvious epistemic virtue either!

In the next section, I will discuss the role of *use* and *similarity* in scientific representation—two central and controversial concepts in this context. Upon that, I will conclude this chapter by summing up why an enhanced version of Giere's account of science is a good basis for our further discussions of interdisciplinarity.

Use & Similarity

In this section, I discuss the concepts of "use" and "similarity" as well as how they interconnect. The issue of similarity is relevant to review in further detail due to the quite heated debate on its role (along with resemblance, likeness, and isomorphism) in representational relations. As Nelson Goodman once wrote:

> The most naïve view of representation might perhaps be put somewhat like this: "*A* represents *B* if and only if *A* appreciably resembles *B*", or "*A* represents *B* to the extent that *A* resembles *B*". Vestiges of this view, with assorted refinements, persist in most writing on representation. Yet more error could hardly be pressed into so short a formula. (Goodman 1976, p. 3 f.)

As briefly mentioned above, an influential argument against similarity is based on the view that while similarity (and isomorphism) are symmetrical relations, representation is asymmetrical. Mauricio Suarez argues, that if *A* is similar to *B*, *B* is also (necessarily) similar to *A*. But if *A* represents

B, B does not (necessarily) represent *A* (Suarez 2003, 2004). As discussed above, Giere has argued in favour of a similarity-based account of scientific representation, though explicitly not of the dyadic kind which Suarez criticises (Giere 2004). Bas van Fraassen, in spite of explicitly expressing his sympathies with Suarez' view, has emphasised that resemblance does actually have a lot to say—at least in some instances of representation. At the same time, he questions the strength of the argument from asymmetry (van Fraassen 2008, p. 17 f.).

There is, I believe, a combination of confusion and disagreement on basic issues at play here. But the difficulties might be quite easily resolved. Van Fraassen's main point with his *Hauptsatz*[9] is, that it is of central importance that representation is understood as an activity and not as a simple (truth-) relation between, for example, linguistic units and aspects of reality. Giere certainly agrees. For *A* to be a model of *B*, someone has to *use A* as a model of *B* (in an act of representation) and thereby indicate that *B* is to be thought of in a certain way. There is, thus, nothing that *in itself* represents something else.

Think of the equation of the simple pendulum discussed in Chap. 1 (and further below). This equation can, given the required specifications, be used as a strongly idealised model of the movements of a pendulum. But a pendulum could also be used as a model of the equation, e.g. as a clarifying exemplification. In any given context, it is exactly the use that determines whether the equation is a model of the pendulum or vice versa. Use is not symmetrical and, consequently, representation (which involves use) is not symmetrical either. This does not preclude similarity relations from being involved in representational relations, though. We use *A* to represent *B*, and as part of this activity we specify in which respects and to what extent we consider *A* and *B* to be similar.

Van Fraassen asks questions such as: "How can an abstract entity, such as a mathematical structure, represent something that is not abstract, something in nature?" (van Fraassen 2008, p. 240). *Use* seems to provide a straightforward solution. The abstract entity does not simply represent the non-abstract target, it is *used* to do so and from this use, it gains the indexicality van Fraassen longs for. This also means that you can use anything to represent anything else. Not all such activities will be fruitful in a scientific sense, of course, but there is nothing in principle that prohibits one from doing so.

In this way, the focus on *use* and *representation as an activity* dissolves another issue that would seriously threaten a similarity-based account of scientific representation. If representation were a question of similarity between a vehicle and a target, one might reasonably wonder what level of similarity would be sufficient for a relation between vehicle and target system to constitute a *scientific* representation. Is a geocentric construal of our solar system, for instance, sufficiently similar to its target to count as a scientific representation of it or is heliocentricity required? Few would probably be prepared to say that Aristotle did not represent the heavenly bodies in his cosmology, or for that matter that his works were not representational in at least a protoscientific sense.

The focus on use dissolves the problem of figuring out what is sufficient for counting as a representation since there are no other minimum requirements than that someone uses *A* to represent *B*. I might actually use my pencil and an open cheese sandwich to illustrate some difficulties concerning the landing of a lunar module on the moon (holes in the cheese might neatly represent some of the deeper craters in which it would be unfortunate to land). This might sound odd at first, but it actually opens up a lot of possibilities, which serve us well in the attempt to analyse the integration of very distant scientific approaches. Especially, it is important that it allows us to include bad representation in the discussion of scientific representation, since we need to be able to figure out if interdisciplinarity may be instrumental in improving poor science. Really bad science, the kind of science one would suppose might benefit the most from interdisciplinary collaborations, would be excluded from consideration if stricter standards of similarity were selected as criteria for being a scientific representation.

Summing Up

The following section will conclude this chapter by adding up the various elements discussed above. This is intended to provide the reader with an overview of what I consider to be the central elements of a workable account of scientific representation. First, though, let me sum up why Giere's position is a good place to start for developing an analytical approach to interdisciplinarity, and how to handle its shortcomings. I am by no means alone in thinking of Giere's account as a great starting point for further developments within the philosophy of science (e.g. Godfrey-Smith 2006, p. 726). However, it seems reasonable to make my personal motivation explicit at this point.

Giere places activities of representation at the very centre of his account, which I believe is *the* way forward when one wants to capture what is common to a very broad-scoped class of disciplines. One of the great challenges in developing a general account of interdisciplinarity is certainly the required scope. A very diverse set of scientific practices may (and do) engage in all sorts of interdisciplinary activities. If an account of interdisciplinarity is to capture general characteristics of the epistemic aspects of this heterogeneous class of scientific activities, it is required with some common ground. A level of analysis is needed that captures something which is both (a) common to all participants and (b) central to substantial epistemic aspects of science. Most existing accounts of interdisciplinarity focus on social aspects and thereby passes on test (a). But they fail on (b). Obviously, most scientific activities involve all sorts of social interaction. This is certainly common ground. On the other hand, it is, as discussed above, far from obvious that a focus on the social will lead to capturing anything substantial about epistemic issues. In this respect, there are reasons to be much more optimistic with respect to approach-based analysis. It is not terribly controversial to claim that all scientific disciplines must involve representation.

It is very central to a characterisation of any scientific activity *how* representation is carried out and *what* is represented. These dimensions are at the heart of Giere's account.

On top of this, and in contrast to at least a good deal of scholars working in the philosophy of scientific representation, Giere explicitly includes the social aspects of science: Who is involved? Who is doing the representation? This is important not least because the social variable is the place to look when one wants to distinguish between e.g. the polymath mode and the social mode discussed above. Further, including social aspects builds bridges to other theorists working within this field. Giere explicitly expresses that he acknowledges the importance of the contributions from the sociology and history of science. Certainly, many contributions from the historical and sociological approaches to science are also highly relevant to the analysis of interdisciplinarity. But it is most fruitful for an epistemic analysis, to focus primarily on representation and having one's representation-based account enriched with social and historical nuances.

Giere emphasises the importance of use. Use is important since it bridges the central representational aspects and the social elements. Further, it dissolves the issue of demarcation which might otherwise threaten attempts at capturing scientific practice.

Giere's account includes purposes which should also be included in the discussion of whether to endorse interdisciplinarity.

One final significant benefit: As pointed out already, when all this is added together we end up having a very close fit between Giere's account of science and the most nuanced account of the nature of disciplines we developed in Chap. 2. This, I believe, more than indicates that we are heading towards fertile ground.

Giere's account, however, has a number of shortcomings as well.

First of all, he fails to make sufficient room for other things than models functioning as vehicles of representation. Even his later more liberal notion of "model" is arguably too narrow for the analysis of interdisciplinarity. Through discussions of various recent accounts of modelling, though, we have arrived at a deflationary understanding of modelling which is sufficiently inclusive for the purpose of analysing interdisciplinarity.

Second, Giere's discussion of the targets represented as simply an aspect of reality is, if not exactly naïve, then at least too restricted to serve our present purposes. This problem is easily handled, since nothing in Giere's account prevents us from including whatever we would like as targets of representation.

Third, the specification Giere provides of the relation between vehicles of representation and targets is inadequate for the present purposes. I have drawn on Thomson-Jones' propositional view of modelling to indicate the roles various propositions play, and on several other accounts, most notably Weisberg, Cartwright, Downes, and Godfrey-Smith, in order to bring out more details about this crucial link. As expressed above, for this purpose I will draw on specifications of approaches, vehicles of representation, and intermediate layers (as constituted by assumptions, algorithms, and tools).

Fourth, Giere does not include outputs. I think it is obvious that an analysis of interdisciplinary activities (as well as all other scientific activities) ought to include some account of the recommendations that are the final output of the scientific activities and, obviously, a central part of the motivation for carrying out research in the first place.

If we combine selected aspects from the early and the later versions of Giere's account of scientific representation with the suggested adaptations, we end up with something like the following: Some person or persons (S) picks out a vehicle of representation (X) by means of a set of statements. (S) uses (X) to represent (W) for some purpose (P) through some more or less explicit intermediate layer. By claiming a specified set of similarities between (X) and (W) some output (O) is produced.[10]

Understanding representation in this way provides us with a number of dimensions along which to search for transformations and differences which could be consequences of interdisciplinarity in the pre- and post-interaction states of affairs.

- (S): Who is involved?
- (X): Which vehicles of representation are involved and are they combined or transformed?
  - How?
- (W): Which targets are involved?
- Which elements constitute the intermediate layer of the integrated approach?
  - How is the use of these elements changed or transformed compared to the parent approaches?
- (P): Which purposes are involved?
- (O): Which outputs are involved?

The most important issue to notice here is that we need to unfold especially the intermediate layer considerably.

The purpose for which we wanted to develop an account of scientific representation in the first place, was to evaluate the epistemic benefits (if any) of cases of interdisciplinarity. The bad (and somewhat old) news is that it is pretty difficult to develop measures for the quality of science (and it is not entirely clear at this stage to what extent my account of scientific representation is helpful in this respect). The good news is that for our present undertaking we do not need absolute measures, but only relative measures. And these are significantly easier to get at. In fact, when viewing the topic in light of the above discussion of modelling and representation, I believe we have a quite decent starting point on the latter enterprise.

In the next chapter, I will move on to discuss some important assumptions about pluralism, idealisation, and distortion prevalent in contemporary philosophy of science (especially in the philosophy of scientific representation). As mentioned above in this chapter, these issues act as double-edged swords within the discussion of interdisciplinarity. On the one hand, they seem to provide all the reasons one needs for devoting oneself completely to uncurbed scientific crossbreeding. On the other hand, they seem to provide reason to think that interdisciplinarity, especially when involving distant approaches, requires considerable more care and effort than many inter-disciples would like to think.

Notes

1. I will include, though, a brief discussion of the consequences of a certain interpretation of Kuhn's concept of "paradigm".
2. Notable theorists associated with these closely related but nevertheless distinct lines of thought are Bas van Fraassen (1980, 1989, 2008), Nancy Cartwright (1983, 1999), Ronald Giere (1988, 1999a, b, 2004, 2006a, b, 2010), Ian Hacking (1983, 1999, 2014), Michael Weisberg (2007, 2012, 2013), Mauricio Suarez (2003, 2004), Anjan Chakravartty (2010), Peter Godfrey-Smith (2003, 2009), Martin Thomson-Jones (2010, 2012), Sandra Mitchell (2003, 2009), and Peter Galison (1997, 2008). Earlier influential discussions in the 1960s and 1970s include work by Nelson Goodman (1976, 1978), Patrick Suppes (1960, 1962, 1967), Frederick Suppe (1972, 1989), and Mary Hesse (1963, 1974).
3. Good places to start, if one is thus inclined, could be (Alspector-Kelly 2004; Hacking 1981).

4. Indeed, in some of van Fraassen's most recent treatments of scientific representation (e.g. 2008) he seems to have developed in a more pragmatic direction as well.
5. Godfrey-Smith credits Deena Skolnick Weisberg for this idea.
6. Giere also mentions words, equations, diagrams, graphs, photographs, and computer generated images (2006b, p. 60).
7. As a word of warning, the derived concept of "model" is not going to overlap nicely with the usage of 'model' among logicians.
8. And of course, not everything that is used as a vehicle of representation is a model in most of these senses.
9. As a service to the reader, I repeat from Chap. 1: "There is no representation except in the sense that some things are used, made, or taken, to represent some things as thus or so" (van Fraassen 2008, p. 23).
10. I have not discussed outputs in this chapter, but I refer the reader to my discussion hereof in Chap. 1.

References

Alspector-Kelly, Marc. 2004. Seeing the Unobservable: Van Fraassen and the Limits of Experience. *Synthese* 140: 331–353.

Boumans, Marcel. 1999. Built-in Justification. In *Models as Mediators— Perspectives on Natural and Social Science*, ed. Mary S. Morgan and Margaret Morrison. Cambridge, UK: Cambridge University Press.

Cartwright, Nancy. 1983. *How the Laws of Physics Lie*. Oxford; New York: Clarendon Press; Oxford University Press.

———. 1999. *The Dappled World: A Study of the Boundaries of Science*. Cambridge; New York: Cambridge University Press.

Chakravartty, Anjan. 2010. Informational Versus Functional Theories of Scientific Representation. *Synthese* 172 (2): 197–213.

Contessa, Gabriele. 2006. Scientific Models, Partial Structures and the New Received View of Theories. *Studies in History and Philosophy of Science* 37 (2): 370–377.

———. 2010. Empiricist Structuralism, Metaphysical Realism, and the Bridging Problem. *Analysis Reviews* 70 (3): 514–524.

Downes, Stephen M. 1992. The Importance of Models in Theorizing: A Deflationary Semantic View. *PSA 1992: Proceedings of the Biennial Meeting of the Philosophy of Science Association* 1: 142–153.

Galison, Peter. 1997. *Image and Logic: A Material Culture of Microphysics*. Chicago; London: University of Chicago Press.

———. 2008. Ten Problems in History and Philosophy of Science. *Isis* 99 (1): 111–124.

Giere, Ronald N. 1988. *Explaining Science: A Cognitive Approach, Science and Its Conceptual Foundations*. Chicago: University of Chicago Press.

———. 1999a. *Science without Laws, Science and Its Conceptual Foundations*. Chicago: University of Chicago Press.

———. 1999b. Using Models to Represent Reality. In *Model-Based Reasoning in Scientific Discovery*, ed. Lorenzo Magnani, Nancy J. Nersessian, and Thagard Paul, 41–57. New York: Kluwer Academic/Plenum Publishers.

———. 2004. How Models are Used to Represent Reality. *Philosophy of Science* 71 (5): 742–752. https://doi.org/10.1086/425063.

———. 2006a. Perspectival Pluralism. In *Scientific Pluralism*, ed. Stephen H. Kellert, Helen Longino, and C. Kenneth Waters, 25–41. Minneapolis, MN: University of Minnesota Press.

———. 2006b. *Scientific Perspectivism*. Chicago: University of Chicago Press.

———. 2008. Models, Metaphysics, and Methodology. In *Nancy Cartwright's Philosophy of Science*, ed. Stephan Hartmann, Carl Hoefer, and Luc Bovens, 123–133. New York; London: Routledge.

———. 2009. Why Scientific Models Should Not Be Regarded as Works of Fiction. In *Fictions in Science: Philosophical Essays on Modeling and Idealization*, ed. Mauricio Suárez, 248–258. New York: Routledge.

———. 2010. An Agent-Based Conception of Models and Scientific Representation. *Synthese* 172 (2): 269–281. https://doi.org/10.1007/S11229-009-9506-Z.

———. 2011. Scientific Models in Philosophy of Science. *Studies in History and Philosophy of Modern Physics* 42 (3): 211–212. https://doi.org/10.1016/J.Shpsb.2011.03.001.

Godfrey-Smith, Peter. 2003. *Theory and Reality: An Introduction to the Philosophy of Science*. Chicago; London: University of Chicago Press.

———. 2006. The Strategy of Model-Based Science. *Biology and Philosophy* 21: 725–740.

———. 2009. Models and Fictions in Science. *Philosophical Studies* 143: 101–116.

Goodman, Nelson. 1976. *Languages of Art: An Approach to a Theory of Symbols*. 2nd ed. Indianapolis: Hackett Pub. Co.

———. 1978. *Ways of Worldmaking*. Indianapolis: Hackett Pub. Co.

Griesemer, James R. 1990. Modeling in the Museum: On the Role of Remnant Models in the Work of Joseph Grinnell. *Biology & Philosophy* 5: 3–36.

Hacking, Ian. 1981. Do We See Through a Microscope? *Pacific Philosophical Quarterly* 62: 305–322.

———. 1983. *Representing and Intervening: Introductory Topics in the Philosophy of Natural Science.* Cambridge, UK; New York: Cambridge University Press.

———. 1999. *The Social Construction of What?* Cambridge, MA: Harvard University Press.

———. 2014. *Why is There Philosophy of Mathematics at All?* Cambridge: Cambridge University Press.

Hesse, Mary B. 1963. *Models and Analogies in Science, Newman History and Philosophy of Science Series.* London; New York: Sheed and Ward.

———. 1974. *The Structure of Scientific Inference.* Berkeley: University of California Press.

Kuhn, Thomas S. 1970. *The Structure of Scientific Revolutions.* International Encyclopedia of Unified Science. Foundations of the Unity of Science, V. 2, No.2. 2nd ed. Chicago: University of Chicago Press.

Mitchell, Sandra D. 2003. *Biological Complexity and Integrative Pluralism.* Cambridge: Cambridge University Press.

———. 2009. *Unsimple Truths: Science, Complexity, and Policy.* Chicago; London: University of Chicago Press.

Morgan, Mary S., and Margaret Morrison, eds. 1999. *Models as Mediators: Perspectives on Natural and Social Science (Ideas in Context).* Cambridge: Cambridge University Press.

Nagel, Ernest. 1961. *The Structure of Science: Problems in the Logic of Scientific Explanation.* London: Routledge & Kegan Paul Ltd.

Suarez, Mauricio. 2003. Scientific Representation: Against Similarity and Isomorphism. *International Studies in the Philosophy of Science* 17: 225–244.

———. 2004. An Inferential Conception of Scientific Representation. *Philosophy of Science* 71 (5): 767–779. https://doi.org/10.1086/421415.

Suppe, Frederick. 1972. What's Wrong with the Received View on the Structure of Scientific Theories? *Philosophy of Science* 39 (1): 1–19.

———. 1989. *The Semantic Conception of Scientific Theories and Scientific Realism.* Chicago: University of Illinois Press.

Suppes, Patrick. 1960. A Comparison of the Meaning and Uses of Models in Mathematics and the Empirical Sciences. *Synthese* 12 (2/3): 287–301.

———. 1962. Models of Data. In *Logic, Methodology and Philosophy of Science: Proceedings of the 1960 International Congress,* Stanford, CA.

————. 1967. What is a Scientific Theory? In *Philosophy of Science Today*, ed. S. Morgenbesser, 55–67. New York: Meridian.

Teller, Paul. 2001. Twilight of the Perfect Model Model. *Erkenntnis* 55 (3): 393–415. https://doi.org/10.1023/A:1013349314515.

Thomson-Jones, Martin. 2010. Missing Systems and the Face Value Practice. *Synthese* 172 (2): 283–299.

————. 2012. Modeling without Mathematics. *Philosophy of Science* 79: 761–772.

van Fraassen, Bas C. 1980. *The Scientific Image, Clarendon Library of Logic and Philosophy*. Oxford; New York: Clarendon Press; Oxford University Press.

————. 1989. *Laws and Symmetry*. Oxford; New York: Clarendon Press; Oxford University Press.

————. 2008. *Scientific Representation: Paradoxes of Perspective*. Oxford; New York: Clarendon Press; Oxford University Press.

Weisberg, Michael. 2007. Who is a Modeler? *British Journal for the Philosophy of Science* 58: 207–233.

————. 2012. Getting Serious about Similarity. *Philosophy of Science* 79: 785–794.

————. 2013. *Simulation and Similarity: Using Models to Understand the World. Oxford Studies in Philosophy of Science*. New York: Oxford University Press.

6

Pluralisms, Perspectives, and Potential Problems

In this chapter, I discuss in more detail some central issues which were only hinted at in Chap. 5, namely the issues of pluralism, perspectives, and distortions. Throughout this chapter, I will draw on Giere's *Perspectival Realism*, the literature on scientific pluralism (Giere 2006a; Kellert et al. 2006; Mitchell 2002, 2012), as well as valuable discussions of perspectives, idealisation, and distortion by van Fraassen (2008), Weisberg (2007), Cartwright (1983, 1999), McMullin (1985), and Wimsatt (2007). Moreover, Godfrey-Smith has provided inspiration for the discussions of *relocational idealisation* below (e.g. 2006, p. 726).

Idealisation is at the heart of the position that representation of a complex world is always in terms of less complex vehicles of representation. Idealisation can be considered a type of epistemic distortion. But there are other types of distortion involved in representation, as well. Central to the arguments of this chapter is that representation always involves distortions—some of which are explicit and deliberate, others of which are unstated and therefore prone to go unnoticed. The natural question is: What happens if one attempts to integrate distorted approaches? Should we expect integration to level out distortions or should we expect some type of (constructive or destructive) interference to occur? Can we

© The Author(s) 2018
R. Hvidtfeldt, *The Structure of Interdisciplinary Science*, New Directions in the Philosophy of Science, https://doi.org/10.1007/978-3-319-90872-4_6

say anything in general about what to expect from divergent distorted interactions?

As this chapter reveals, once one starts looking for it, one quickly finds a pluralism of pluralisms within science. Pluralisms regarding perspectives, distortions, and ceteris paribus clauses. Internal, external, metaphysical, and epistemic pluralisms. And, not least, a pluralism of reasons for being cautious when integrating scientific approaches. As the following will show, maintaining a (rather plausible) weak pluralistic position coupled with a mild form of realism (which is rather plausible as well, I believe) spells trouble for interdisciplinarity. There might be ways out of this predicament, but certainly one should carefully consider whether relying on easy short-cuts is the way to go.

Pluralism—What is It?

Quite appropriately, the answer to the question "what is scientific pluralism?" is multi-dimensional. There is, in other words, a multitude of ways in which science is pluralistic, at least according to advocates of scientific pluralism. Central, however, to most pluralistic accounts of science is a monistic assumption, which one might call "the one-world-hypothesis".

Giere has phrased the one-world-hypothesis in the following terms:

> Imagine the universe as having a definite structure, but exceedingly complex, so complex that no models humans can devise could ever capture more than limited aspects of the total complexity. Nevertheless, some ways of constructing models of the world do provide resources for capturing some aspects of the world more or less well. Other aspects may provide resources for capturing other aspects more or less well. Both ways, however, may capture some aspects of reality and thus be candidates for a realistic understanding of the world. [...] It does not matter that different historical paths might lead to different sciences. Each might genuinely capture some aspects of reality. (Giere 1999, p. 79)

Giere's view is that different scientific models emphasise different aspects of reality. Reality itself is so immensely complex that no scientific model (or theory) will ever be able to capture all of its aspects. Giere emphasises that he intends the one-world-hypothesis to be understood as an instrumentally motivated working hypothesis—not as a piece of a priori metaphysics (Ibid.).

The hypothesis that there is one, and only one, reality may seem self-evidently true (especially if one is realistically inclined). After all, if there were more than one world or reality, one might reasonably argue that all these different realities would add up to the sum total of realities, which could then simply be called 'reality'. But, at least in principle, there are other possibilities. Especially if considered in less realistic perspectives, matters sometimes appear less straightforward than described by Giere.

Some other possible positions are:

1. There does not exist a mind-independent reality.
2. There exists a plurality of distinct mind-dependent realities.
3. There exists a plurality of distinct, non-interacting mind-independent realities.
4. Reality consists of one (or more) physical mind-independent reality(s) plus a number of other kinds of reality contingent on, for example, social and mental factors.

One could undoubtedly find people actually endorsing each of these positions, even if none of them seem very attractive. But even though there are indeed an abundance of writings on the social construction of various things,[1] it is quite rare to encounter people who seriously doubt that physical reality exists independently of someone's mental or social activities.[2] Position (4), on the other hand, is not all that implausible as long as the different realities are not thought of as isolated, non-interacting domains. But in that case, (4) would not be incompatible with the one-world-hypothesis.

In *Explaining Science*, Giere discusses the views of Gilbert and Mulkay, in *Opening Pandora's Box* (1984), as an example of someone apparently believing in multiple realities. Indeed, the opening line of the final chapter of their book is: "In this book we have approached the social world of science as a multiple reality" (ibid., p. 188). But notice that it is the *social* world of science they have approached as a multiple reality. Not the physical world, which the science in question targets.[3] Gilbert and Mulkay's main conclusion is that scientists operate with distinct frameworks for explaining their own results and other scientists' conflicting results. Accordingly, scientists often construe their own results as determined strictly by experimental data, whereas contingency and the influence of social factors is emphasised in descriptions of the activities of opponents. Even though Gilbert and Mulkay's topic is certainly interesting and the reality of the problems they address is plausible, it seems sufficient to talk of multiple accounts or frameworks to explain the phenomena in question (Giere 1988, p. 61). There is no reason to postulate distinct realities.

Sandra Mitchell has suggested an alternative: "However complex, and however many contributing causes participated, there is only one causal history that, in fact, has generated a phenomenon to be explained" (Mitchell 2002, p. 66). This would be true regardless of whether there are indeed multiple realities, and whether or not these multiple realities were capable of interacting. One might call this position causal realism involving a subscription to a one-causal-history-hypothesis.

It is difficult to think of good arguments against the one-world-hypothesis or the one-causal-history-hypothesis. Indeed, these seem to constitute a reasonable default position when pursuing an analysis of interdisciplinarity. So for now, let us allow ourselves to be monists regarding the causal structure of the world.

Of course, Mitchell's account only holds when considering *tokens* of a given phenomenon. If we instead consider a *type* of phenomena, the tokens of any type will have divergent causal histories—sometimes even qualitatively quite different causal histories. One might argue that, at least in some cases, causal history has a lot to say when deciding whether two distinct phenomena are of the same type. But certainly, this cannot be generalised. There are lots of phenomena which are categorised as being tokens of some type in complete disregard of their causal history. A house is a house regardless of how it was built. "God only knows" and "Heroes" both belong to the category "love songs performed by David Bowie", even though the first (tormentingly embarrassing cover version) came into existence as a consequence of bad career management, while the latter is considered the result of a musical genius at his creative peak. Causal history has little influence on categorisation in many cases. This will, for instance, turn out to be crucial in the psychopathological cases discussed below.

Perhaps, in the above quotation, Giere comes across as a little rash in writing off concerns about there being different and competing sciences. Mitchell, at least, refers to this as perhaps the chief predicament of a pluralistic realism:

> [I]f science is representing and explaining the structure of the *one* world, why is there such a diversity of representations and explanations in some domains? One response is that pluralism simply reflects the immaturity of the science (Kuhn 1962). Yet history shows us that many sciences do not exhibit a diminution in the multiplicity of theories, models, and explanations they generate. This 'fact' of pluralism, on the face of it, seems to be correlated not with maturity of the discipline, but with the complexity of the subject matter. Thus the diversity of views found in contemporary science is not an embarrassment or sign of failure, but rather the product of scientists doing what they must do to produce effective science. Pluralism reflects complexity. (Mitchell 2002, p. 55; reference to Kuhn in the original)

The question seems to be whether one can coherently combine the following three elements into a coherent position:

1. Support for the one-world-hypothesis.
2. Belief in the ability of science to generate (approximately) true representations of reality.
3. Acknowledgement that there is a multitude of different and often conflicting ways of representing and explaining even apparently simple phenomena.

I take the above three elements to constitute the core of pluralistic realism. As I have stated clearly above, I have no ambition of entering into the realism vs. anti-realism-debate in this book. But it is worth mentioning that in light of the above, anti-realism is obviously less problematic to defend. The anti-realist may respond that the plurality of models and theories is simply the best method of optimising empirical adequacy. The realist, on the other hand, is obliged to deliver a more elaborate response to this challenge (Weisberg 2007, p. 656). We can uncover some central parts of such a response by considering different dimensions of scientific pluralism in more detail.

The Pluralisms

To make sense of the tension between the one-world-hypothesis and conflicts at the theoretical level, we need to disentangle the multitude of
pluralisms at play within the scientific pluralism position. The following
will reveal the most important dimensions along which science is pluralistic. Some of these dimensions will be recognisable from discussions in
previous chapters.

The following section will briefly discuss four ways in which science is
pluralistic. Of the four types of pluralism, I believe we are ready to accept
the first two (internal and external pluralisms) on grounds of evidence
provided above. These are, certainly, both important in the present context. Regarding the third type, metaphysical (or nomological) pluralism,
I will remain agnostic, since it is less important for present purposes. For
present purposes, there is more than enough to deal with when considering the more accessible ways in which science is pluralistic.

The fourth type of pluralism discussed in the following, epistemic pluralism, is the central component of Giere's perspectivism as well as a very
central component in a representation-based discussion of interdisciplinarity. This, then, is the most important addition to my account in the
present chapter. Therefore, I will embark on a more thorough discussion
of epistemic pluralism and perspectives after the following (brief) discussion of the first three dimensions of pluralisms.

#1—Internal Pluralism

Internal pluralism refers to the diversity of the class of elements involved
in theorising and modelling. Examples, as discussed in the previous chapter, include mathematical and meta-mathematical tools and concepts,
propositional algorithms, various (e.g. ontological or epistemic) assumptions, ways of arriving at distinctions and classifications, all kinds of
kinds, concepts, metaphors, analogies, values, and policy related issues.

All of these are candidates for being transferred as part of interdisciplinary activities. Further, none of these are guaranteed to work well in
new contexts simply because of having performed satisfactorily in the

contexts from which they are imported (this is *insight a* again). I will provide detailed examples hereof below in this chapter.

#2—External Pluralism

External pluralism refers to the many different *kinds* of models and theories. We have already discussed the prevalence in the literature of discussions of physical models, diagrammatic models, historical models, remnant models, mathematical models, fictional models, and model organisms (Downes 1992; Griesemer 1990; Levy and Currie 2014; Weisberg 2013). Further, computer simulation is on some accounts considered to be a distinct type of modelling for being situated somewhere between theory and experiment and for incorporating practical and epistemological challenges entirely different from other types of modelling (Lenhard et al. 2010; Weisberg 2013; Winsberg 2010).

It is certainly possible to transfer types of models as part of an interdisciplinary activity.

The diversity captured by the internal and external dimensions of pluralism is central to the philosophical opposition to accounts of modelling along strictly mathematical and/or set-theoretical lines discussed in the previous chapter (Downes 1992; Godfrey-Smith 2006; Thomson-Jones 2012; Weisberg 2013).

#3—Metaphysical (Nomological) Pluralism and CP-Clauses

Nancy Cartwright's work on modelling, pluralism, and the limitations of scientific theories is among the most influential contributions of "The Stanford School" (Cartwright 1983, 1989, 1999; Morgan and Morrison 1999). In (1999) Cartwright discussed metaphysical pluralism, which she characterised as "the doctrine that nature is governed in different domains by different systems of laws not necessarily related to each other in any systematic or uniform way; by a patchwork of laws" (p. 32). In case metaphysical pluralism reflects reality, it would certainly have consequences for the viability of transferring theoretical elements from one context to another. Indeed, the better the fit between model and its target, the less likely it might be for the model (or some other element involved in modelling) to fit well in other contexts. For now, however, I will remain agnostic on this issue, partly due to the difficulties related to settling the matter, and partly because (as already mentioned) the other three pluralisms (#1, #2, #4) provide us with more than sufficient material to consider for present purposes.

Cartwright is the major proponent of the idea of metaphysical pluralism, but she has also contributed significantly to developing insights into other dimensions of modelling and theory construction, including the internal/external distinction. As part of her discussion of these issues in 1999, Cartwright made quite a convincing case that the utility of theories and models is strongly restricted by ceteris paribus clauses. Laws, models, and theories work well in many cases, she states, but these many well-functioning cases are outnumbered (by far) by the many cases in which laws, models, and theories do not work well (if at all). And when something works well, it only works well ceteris paribus. Therefore, there is little reason to assume that a given model or theory will work well if conditions, or contexts, change.

I think we have good reasons to assume that something similar is the case with the internal elements as well as the external kinds of modelling. We learn from internal and external pluralism that when we represent reality, we make use of many different kinds of elements, which we combine in many different ways. These elements, and the different ways in which they are combined, are developed for use in certain contexts, and their performance has been calibrated and evaluated under specific, context-dependent conditions. Even if they work well in a particular context, their satisfactory performance cannot be generalised.

A number of dimensions have been distinguished along which generalisations might be non-universal:

> [...] many central and well-entrenched generalizations in the sciences deviate from the received characteristics by being "non-strict", "inexact", "exception-ridden", "contingent on the circumstances", "sensitive", "non-robust", "idealized", "abstract", "merely statistical", and so forth. We call generalizations with these features "non-universal". (Reutlinger and Unterhuber 2014, p. 1705)

Non-universality is multifaceted. This means that sensitivity to changes in, for example, idealisations, use of statistical tools, and initial and background conditions may differ from case to case. Viewed in this perspective interdisciplinarity takes on a distinctly audacious, perhaps even reckless, flavour. When you integrate or combine approaches or apply vehicles to new targets, other things certainly aren't equal.

It is superfluous to state that one should be careful when generalising. In this context, though, it is perhaps fair to revitalise this commonplace by stating that (generally) one should pay attention to potential risks related to ignoring ceteris paribus clauses when transferring theoretical elements between contexts. If anything is characteristic of interdisciplinarity in general, however, it is not, I believe, close attention to ceteris paribus clauses. In this chapter and the next, I will provide examples of tools and assumptions that seem to work well under certain conditions, and less well when conditions are different. These examples will illustrate that not taking these matters into account might have quite serious consequences. Further, in relation to the discussion of relocational idealisation below in this chapter, I assert that ceteris paribus clauses may be lost when vehicles of representation, or other theoretical elements, are transferred between contexts.

In the next section, however, I will turn to what I term 'epistemic pluralism' which is very central to the accounts of, among others, Giere and Mitchell. This dimension of pluralism adds considerably to our growing appreciation of the difficulties related to interdisciplinarity.

#4—Epistemic (Representational) Pluralism

For present concerns, the most interesting pluralism is what one might call *epistemic pluralism*, which indicates that there is a multitude of ways to know something about any given phenomena. Importantly, several ways of knowing something about some object may all be useful, even though they seem to be mutually exclusive. Therefore even seemingly incompatible models can supplement each other in useful ways (Giere 1999, p. 79). An example could be two different models of water, one in which water is considered a fluid, the other in which water is considered a collection of molecules. The first model is good for studying the viscosity of water, but if you are interested in Brownian motion, the second is better.

If we compare two other classical examples of competitive accounts, interesting questions arise. Compare the (apparently) conflicting wave and particle theories of light with the (apparently) conflicting Darwinian and Lamarckian theories of inheritance. In (2002) Mitchell discussed these as two instances of what she calls competitive pluralism. According to Mitchell, competitive pluralism is characterised by the presumption that pluralism is temporary and will eventually be eliminated. This she distinguishes from compatible pluralism, that is a pluralism of models which are not mutually exclusive (even though they cannot be unified) and can therefore be expected to co-exist peacefully.

It seems, though, that we need a further distinction within the group of theories and models subsumed under competitive pluralism. At least considered at a certain level of generalisation, we need to ascribe as well particle as wave properties to light in order to represent the phenomena we observe. This is not the case (at least not according to established common scientific sense) with the competing theories of inheritance. In a sense you can say that in the case of light, *the phenomenon requires multiple representations* which are (at least at the face of it) incompatible. The same is not the case regarding the inheritance issue, in respect to which Mendelian genetics simply undermined the trust in inheritance of acquired traits.

The central claim of epistemic pluralism is that different ways of knowing something about some topic may not be mutually exclusive in the sense that we will, eventually, have to decide upon one or the other. On the pluralistic account, they may both be required for our best combined explanation even though they appear to conflict.

Perspectivism

Giere's discussion of perspectives is a nicely elaborated example of epistemic pluralism (the most central text is 2006b). A central assumption in Giere's *Perspectival Realism* is that representation always involves a certain perspective. The metaphor of perspectives is both quite rich and quite dominant in Giere's later work. Therefore, it is worth taking a close look at—not least since a focus on perspectives has important consequences for any epistemic account of interdisciplinarity. As in all cases of metaphorical thinking, though, one must be careful not to put too much weight on the literal meaning of 'perspective'. Giere uses 'perspective' as a technical term that captures central aspects of literal visual perspectives involved in observation, as well as important issues about how the internal make up of theories, tools, and models affect the results we get from applying them.

On Giere's account, scientific representations are perspectival in a number of different ways. The most straightforward example is at the level of observation. An accessible illustration is the views of some object, for instance a house, from different positions. Straightforwardly, any way of observing a house will implicate a perspective: You see the house from above, from its southeast corner, or from the inside, and so on. There is no view from nowhere and neither a view from everywhere which can provide all perspectives simultaneously—any observational perspective shows only part of the phenomenon in focus (Giere 1999, p. 79 f.).

The physiology of our sense organs adds another dimension to the metaphor. If I look at a house, I experience it in colour. Someone with monochromatic vision would experience it quite differently. Put in very simple terms, Giere considers this human ability to see colour as a perspective (Giere 2006b, p. 17 ff.). And the specific perspective depends on how inputs from the environment are processed (ibid., p. 59). The perspective of normal colour vision emerges when a specific set of physiological circumstances are in place and are fully operational.

As Giere states:

[...] [M]aybe the most important feature of perspectives is that they are always partial. When looking out at a scene, a typical human trichromat is visually affected by only a narrow range of all that electromagnetic radiation available. (Giere 2006b, p. 35)

Thus, only parts of the available phenomena are registered during observation due to limitations inherent in the physical makeup of normal vision.

Importantly, light with distinct spectral characteristics, which are experienced as identical by the person with dichromatic colour vision, are in some cases experienced as clearly distinct by the person with normal trichromatic colour vision. A tetrachromat, that is, a person whose retinae contains four classes of cones instead of the usual three, is further capable of discriminating spectral stimuli indistinguishable to the person with normal colour vision (Jordan et al. 2010). Phenomena may, then, when viewed in one colour perspective, come across as obviously identical,[4] but as clearly distinct when viewed in a different perspective.

We may add further dimensions to our understanding of observational perspectives by considering scientific equipment. If we observe the house through an electron microscope, we get an entirely different observational perspective. Each of these different observational perspectives provide access to reality, but they all draw only a partial image hereof.

As Giere states, scientific instruments

> [...] process inputs from the environment in ways peculiar to their own physical make up, ways that render these inputs similar or different not just according to features of the inputs themselves, but also according to features of the instrument. (Giere 2006b, p. 59)

The same is certainly true about the human visual system.[5] I will return to, and paraphrase, this quotation several times below. It captures something very central which also applies to the perspectives of models, theories, conceptual tools, and approaches. I will demonstrate the significance of this in discussions of Simpson's paradox and operational definition below in this chapter.

Importantly, these different perspectives cannot be integrated in a unified "inter-perspective" without (further) distorting the individual contributions. We cannot, for instance, visually represent a house from without and within in one and the same image without significantly distorting its appearance to the naked eye (Giere 2006b, p. 14). Further, it is certainly hard to imagine how it should be possible to integrate the perspective of monochromacy with that of trichromatic ("normal") colour vision (not to mention dichromatic, tetrachromatic or plain achromatic colour vision).

Perspectives of Theory

The next step is to explain the perspectives of scientific theories and models. This step forces us further away from the literal meaning of 'perspective'. By analogy, Giere claims that theories represent reality in the same way that maps represent (parts of) the world. A map can represent in many ways: quantitatively or qualitatively, in more or less detail, focusing on various aspects (e.g. geographical, political, geological, demographical), and so on. This corresponds to the different perspectives of models selectively emphasising aspects of the world.

Map projection is a neat example one might offer as an illustration of Giere's analogy. Map projection is the construction of two-dimensional representations of curved three-dimensional objects. The most obvious example is the representation of the Earth on a plane map. The Earth has many properties which are candidates for representation on maps, for example, area, distance, shape, direction, and so on. Map projections can be constructed to preserve some of these properties, but not all of them at once. Any projection will compromise or approximate basic metric properties.[6] And as a consequence, every distinct map projection distorts reality in some way. Large scale maps of small proportions of the Earth may come close to being accurate, but distortions build up as the size of the depicted area increases.

As opposed to (some) observations, maps, like theories and models, are deliberate constructions. They are vehicles of representation constructed for specific representational purposes. Different ways of drawing maps emphasise different aspects of reality. One can even construct maps that deliberately misrepresent, or maps that represent fictive places. Further, properties of a map can be re-projected onto reality (think, for instance, of the global positioning system). This reciprocal relationship is also a feature of the relationship between scientific theories and reality.

To make the move from analogy to concrete examples of theoretical models, Giere uses the mathematical description of the movements of a simple pendulum.[7]

$$p = 2\pi \sqrt{l/g}$$

Fig. 6.1 Return of the simple pendulum

Again: In Fig. 6.1, P is the period (the length of a full cycle) of the pendulum, g is the local gravitational acceleration and l is the length of the pendulum. The equation corresponds to experimental results by Newton and Galilei, which showed that a pendulum's period is proportional to the square root of its length and independent of its mass. This equation is far from an accurate description of how a physical pendulum actually moves, though.

The equation corresponds to the motion of a simple pendulum, which is an idealisation of a real pendulum based on the following assumptions: The bob is a point mass; the rod or cord on which the bob swings is massless, inextensible, and remains taut at all times; motion is frictionless; the angle of the swing is very small, so that the two-dimensional swing of the actual pendulum approximates a strictly horizontal, one-dimensional motion (Giere 1988, p. 70).

On this background, we can safely conclude that the mathematical model of the simple pendulum is quite strongly idealised. Point masses and massless suspensions are prototypical instances of what in the literature are called Galilean idealisations (McMullin 1985). No such things exist in physical reality. The equation is further idealised since it does not include aspects such as the materials of which the pendulum is made and the climatic conditions where it is placed (this is the Aristotelian sense of idealisation[8]). The choice of material of the bob does, however, matter in relation to how much the movement of an actual pendulum is affected by the earth's magnetic field, while the length of the suspension may vary as a result of changes in temperature (to different degrees depending on what it is made from). Moreover, this model in no way takes into account the moving pendulum's reflection of light, its history (e.g. the role played by pendulum movement in the determination of the standard meter), or its potential hypnotic effect. Consequently, one must conclude that the equation is very far from an exhaustive representation of the movements of physical pendulums. No one, of course, claims that it is anything but a highly idealised mathematical model.

A model that excludes friction, mass, extension, and so on, and is only able to deal with a small fraction of the pendulums actual potential amplitude can reasonably be termed "partial". But notice, then, how insight into these idealisations are required in order to be able to make proper use of the equation to represent a pendulum. In other words: The equation is worth little as a vehicle of representation if isolated from the assumptions and algorithms by which it is connected it to its (traditional) target. Knowledge of the intermediate layer is required to use this (and any other) specific vehicle of representation appropriately.

How is the equation of the simple pendulum perspectival? By only providing a partial image of its target (by means of idealisations) and by processing inputs in ways peculiar to its own conceptual makeup. To completely appreciate the perspectives of vehicles of representation, however, we need to construe them as part of the approach in which they are used.

In this context, an approach is defined as a specific way of using a specific vehicle of representation to represent a specific (kind of) target. A specific way of use incorporates the conceptual tools, procedures, and assumptions which serve to link (the selected parts of) the vehicle of representation to (the selected parts of) the target.

The conceptual make up of an approach is constituted by a specific combination of the internal elements that are used in the context. If an element is changed, so is the perspective (and the approach). This is to be understood in analogy to the different perspectives of vision where differences in physiological constitution are used to account for different observational perspectives.

Against this background, we can rephrase Giere's statement about scientific instruments in the following way:

> An approach processes inputs in ways peculiar to its own conceptual make up, ways that render these inputs similar or different not just according to features of the inputs themselves, but also according to features of the approach.

Laws and Perspectives

According to Giere, then, even laws are perspectival. Indeed, Giere understands laws as defining basic perspectives:

> [...] the grand principles objectivists cite as universal laws of nature are better understood as defining highly generalized models that characterize a theoretical perspective. Thus, Newton's laws characterize the classical mechanical perspective; Maxwell's laws characterize the classical electromagnetic perspective; the Schrödinger equation characterizes a quantum mechanical perspective; the principles of natural selection characterize an evolutionary perspective, and so on. On this account, general principles by themselves make no claims about the world, but more specific models constructed in accordance with the principles can be used to make claims about specific aspects of the world. And these claims can be tested against various instrumental perspectives. Nevertheless, all theoretical claims remain perspectival in that they apply only to aspects of the world and then, partly because they apply only to some aspects of the world, never with complete precision. (Giere 2006b, p. 14 f.)

Below, I will expand Giere's notion of perspectives to further include conceptual tools in addition to observation, instruments, models, theories, approaches, and laws. This is required for approach-based analysis to function. But first I will turn to the discussion of some consequences of partiality and perspectivism.

Distortions

Recent discussions of scientific perspectives (Giere 2006b; van Fraassen 2008) highlight how the selectiveness of representation results in various forms of distortion. In the words of van Fraassen: "It seems then that distortion, infidelity, lack of resemblance in some respect, may in general be crucial to the success of a representation" (van Fraassen 2008, p. 13). In the following, I will provide some examples of inevitable representational distortion.

Distortion is a central and important topic in itself in discussions of scientific representation. However, issues related to distortion take on a whole new level of significance in cases of interdisciplinary integration. Interdisciplinarity, thus, does not neutralise van Fraassen's verdict quoted above. Interdisciplinary representation will still involve distortion. But distortions may be transformed as a consequence of integration, and these changes may be quite difficult to track. These questions are important, since it would be naïve to assume that distortions level each other out when approaches are integrated. Only quite careful analyses might reveal the consequences of integrating distortions.

Idealisation

As already touched upon, idealisation is a central type of distortion discussed at length in the philosophical literature. Especially Cartwright (1989), McMullin (1985), Wimsatt (2007), and Suppe (1972) have contributed to establishing the centrality of idealisation in philosophy of science. Idealisation is often characterised as deliberate misrepresentation, either by distorting representational elements or entirely leaving out some of the elements known to actually be a part of the target system. However, it is worth keeping in mind that in many cases misrepresentations may not be deliberate.[9] In some cases it may not be clear whether or not, or to what extent, central elements are indeed misrepresented. Especially relevant to cases of interdisciplinary integration, awareness that some element is misrepresented may be reduced when elements are handled in settings with less than optimal expertise.

One might think of idealisation as the theoretical (modelling) counterpart to the experimental isolation of causes and effects. In *Nature's Capacities and their Measurement* (1989), Cartwright indeed states that abstraction (her preferred term for Aristotelian idealisation) is the process of subtracting everything but the *causal* factors from a model. In experimental settings, of course, you will often get an unmistakable response if you isolate in inappropriate ways. That is, the effect you are attempting to control or generate will change or disappear if you remove central, contributing causes. One should not expect such clear feedback when idealising beyond the reasonable. In principle, nothing prevents the postulation of causal relationships between any elements one might think to combine in an approach. The lack of response is especially treacherous in cases where it is not feasible to construct empirical tests or simulations of the situations in question, for one reason or another. Further, as argued above, there is nothing that stops us from developing non-causal models which are, indeed, also idealised in all sorts of way. Consequently, it will be inadequate to focus exclusively on capturing causal elements when discussing idealisation.

One overall goal of representational activity is to emphasise certain features, to make some point, or show that some aspect is particularly significant. When we represent a person as suffering from schizophrenia, we are not interested in his or her digestion. Therefore, processes of digestion are excluded from most models of schizophrenia, because we want to focus only on issues we take to be relevant to the phenomena we wish to understand. This is only partly due to concerns for cognitive economy. Incidentally, there is a widespread agreement that matters of digestion does not play a vital part in the aetiology of schizophrenia. Nevertheless, the idealisation involved in excluding digestion from consideration results in a somewhat distorted representation, since suffering from schizophrenia actually does involve digestive processes (in the sense that digestion is necessary for staying alive, and dead people do not suffer from schizophrenia, at least not according to the operational criteria of current diagnostics).

Above, we discussed the idealisations involved when representing by means of the equation of the simple pendulum. An examination of a quite different model will show that the characteristics of representational idealisations are repeated across traditional disciplinary boundaries, with some significant differences though.

The *diathesis-stress model* is an example picked from psychopathology. The model is used to cast light on why some people develop pathological mental disorders in contexts which others are able to cope with without similar consequences (Ingram and Luxton 2005).[10]

'Diathesis' here refers to a person's level of vulnerability and may be understood as the opposite of resilience. 'Stress' is not to be understood in the everyday sense, but as any major or minor life event that disturbs the stability of a person's physical, emotional, or cognitive mechanisms. In the simplest version, the diathesis-stress model has only two parameters, vulnerability and accumulated stress, which, taken together, determines if the person suffers "a breakdown" and develops a mental disorder.

Graphically, the diathesis-stress model can be represented as in Fig. 6.2. Much like the equation for the movements of a pendulum the diathesis-stress model is strongly idealised in the Galileian sense. The model is focused upon an alleged isolatable and generalisable aspect of reality, that is, the relation between the tendency of human beings to develop mental disorders and their exposure to stressors. The two parameters, "diathesis" and "stress", are also strongly idealised. Vulnerability seems to be a very dynamic and complex phenomenon, quite far from being a static innate threshold. Similarly, stress can hardly be claimed to accumulate in any simple way until "critical mass" has been reached and a mental breakdown occurs.

The breaking of a leg can, of course, be a significant stressor, but it seems reasonable to claim that the resulting stress decreases as physical functionality is regained. Similarly, the stressfulness of more emotionally traumatic events, such as the divorce of one's parents or the loss of a loved

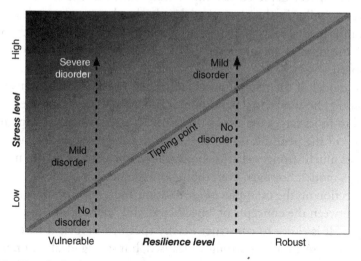

Fig. 6.2 The diathesis-stress model

one, tend to decrease over time. And even though one might develop a persistent hyper-sensitivity in situations reminiscent of the one in which the original crisis occurred, this by no means indicates a simple accumulation of stress over time. Considerations about the decay of stressors are relevant unless one interprets the model so radically as to only include stressors which coincide exactly. But this would make the model more or less useless.

Furthermore, the model is idealised in the (Aristotelian) sense since several (relevant?) factors are left out of the equation (so to speak). Helpful social relations might significantly reduce the impact of a given stressor; so might the acquisition of coping-strategies. And, as the old wisdom goes, overcoming stressful situations might actually strengthen one's ability to combat novel trials.

The diathesis-stress model and the equation of the simple pendulum are two very different models. They nevertheless share certain patterns, which render them comparable as potential vehicles for representation. The main difference between the two is that the equation of the simple pendulum, obviously, has a mathematical structure doing a significant part of the work. The diathesis-stress model, on the other hand, involves no apparent mathematical tools (though it is sometimes, as above, somewhat deceptively depicted as a linear function). But the two models are also very similar in the sense that they both rely on more or less explicated intermediate layers, which serve to point out which aspects of the world the models are supposed to be about, how to properly idealise, and so on.

The model of the pendulum entails much more specific claims about (aspects of) how an actual pendulum is supposed to move in order for the model to fit its target than does the diathesis-stress model. We might say that use of the pendulum-model is a stronger representational act, since it makes much stronger claims about reality. The diathesis-stress model is no less a model, however, even though it is somewhat feeble.

Reflection on the examples of idealisation above reveals important tensions between the concepts of "specificity" and "explication" discussed in Chap. 5. It is easier to construct exactly specified models if its elements and internal relations are strongly idealised, but strong idealisation makes it more demanding to explicate the relation between the model and the phenomena it is about.

In the ordinary use of the model of the simple pendulum it is clearly specified what its elements are and which parts of which phenomena they are used to represent. Further, it is very explicit which relations are taken to hold between the elements of the model. This specificity, however, very much depends on the high degree of idealisation of the model's internal relations and elements. In the case of the diathesis-stress model the elements of the model and the relations between them are less clearly specified and explicated, though still highly idealised. It is worth noting that the literature on simple pendulums is also very explicit about the idealisations used in the model (Giere 1988, p. 69 f.). This is unfortunately not the case in the treatment of diathesis-stress models, in which the transformation from real world phenomena to idealised element of the model is left in the dark (Ingram and Luxton 2005). Consequently, the model of the simple pendulum is a High-Fidelity model (as discussed in Chap. 5). In contrast, the diathesis-stress model is very loosely described, which, among other problems, makes it difficult conclusively to decide whether the model is empirically adequate. In Weisberg's terms, we might simply be unable to determine whether the model lives up to its fidelity criteria (if such had been worked out in the first place).

Relocational Idealisation

When tools, algorithms, assumptions, or vehicles are transferred from one setting to another it is often the case that the transferred elements are altered during the process. One might think of this in terms of *idealisation of theoretical elements*. Both types of idealisation discussed above apply here. Sometimes certain parts of the theoretical structures are considered to be essential and only these are transferred (the Aristotelian approach to idealisation). In other cases, simplified and otherwise distorted versions of theoretical elements are put to use in new settings, in analogy with the Galilean approach to idealisation.

I suggest thinking of idealisation of theoretical elements transferred between representational contexts as *relocational idealisation*. Explicating whether and how relocated theoretical elements are idealised during integration of approaches is a central and important task when analysing cases of interdisciplinarity.

I will provide examples of relocation and relocational idealisation below in this chapter.

Approximation

Approximation is another kind of distortion often distinguished from idealisation, though they are closely related. One might distinguish between numerical, algorithmic, conceptual, and empirical approximations.

Numerical approximations are straightforward: "Every case in which a quantitative property of an object in the system is substituted in the model by a property that has an approximately similar value is a case of approximation" (Contessa 2006, p. 374).

The discussion of the statistical tool ANOVA below will provide a detailed account of an algorithmic approximation, in which a simpler algorithmic structure is used instead of a more accurate, but also more complex, structure. The use of such approximations has been shown to cause significant problems in many cases.

With respect to conceptual approximations, one might think of the non-expert's use of more general categories as a form of approximation. An interesting study of this phenomena involves the comparison of folk-biological categories and classifications among tree-experts and the influence of these differences on inductive reasoning (Medin et al. 1997). Differences in terms of fundamental conceptual structure are not automatically imported along with models or tools in interdisciplinary integration. This may cause confusion and be a source for unintentional distortion.

An example of what I refer to as empirical approximation is described in the following quotation by Deena Weisberg:

> All brains are shaped and organized slightly differently, just like other parts of the body. My brain might be slightly smaller than yours, or my hippocampus located slightly more to the left. This means that a scan of

my brain and a scan of your brain would not overlap exactly. But research studies require responses from multiple participants to ensure that the phenomenon under study is general, not subject-dependent. To solve the difficult problem of comparing the spatial structure of many brains when each of these structures is different, scientists have developed technical methods for standardizing each brain picture to fit a common template. This process ensures that all of the brains to be compared are of exactly the same shape and size, allowing the creation of a single measure of brain activation from brains with disparate shapes and sizes. (Weisberg 2008, p. 52)

Such approximations are certainly not always emphasised in publications of interdisciplinary results informed by neurobiology. When elements are relocated between approaches, appreciation of all kinds of approximation is likely to be idealised away.

Distortions of Scale

Specific challenges are involved in the construction of scaled models. An example of a scaled model could be a 30 × 30 × 12 centimetres model of a 10 × 10 × 4 metres wooden cabin. Scaled models involve important (and to some extent unapparent) types of distortion. The diminished wooden cabin is necessarily distorted since not all properties of wood changes proportionally when they are upscaled or downscaled. The weight of a cylindrical beam is proportional to its volume, which is again proportional to the cube of its radius. But the strength of the beam de- and increases proportionally to the square of its radius, why, obviously, the mass to strength ratio will not remain constant even through perfect geometrical up- and downscaling (van Fraassen 2008, p. 49 f.).

This example is banal by the standards of contemporary engineering. Nevertheless, it is an example in which it is impossible to create a non-distorted scale model. One cannot simultaneously retain the geometric proportions and the mass to strength-ratio. Up- and downscaling along various other dimensions will often involve more or less apparent distortions. One must choose between different distortions, and appropriate choices based on the characteristics of a given model can only be made if the adequate insights are available. When integrating elements from disciplines distant from ones core expertise, insights into technical details tend to be shallow.

Simpson's Distortions

As a further example let us look at a case of statistical distortion, which is, as I shall discuss below, especially relevant to analyses of contemporary psychopathology. Simpson's paradox is named after the statistician Edward H. Simpson who pointed out a quite counterintuitive aspect of associations between fractions and potential consequences of sub-division (Simpson 1951).

Let us assume the following scenario: A medical company wants to test the effect of a wonderful newly developed antidepressant (with no side-effects, of course). The gold standard of such tests is the randomised double-blinded study: N number of participants are randomly divided into two groups receiving the new drug and a placebo respectively. In the following example (Fig. 6.3; $N = 878$), the data shows the (statistically significant) result that treatment (T) has a positive effect on recovery from the treated disorder (D):[11]

	R	~R	rec.-rate
Treatment	369	340	(≈52%)
Placebo	152	176	(≈46%)

Fig. 6.3 Positive effect of treatment

If the population is partitioned by gender (as in Fig. 6.4), however, we get the opposite result: T has a negative effect on recovery from D as well in the male part of the population as in the female part of the population (this effect is also statistically significant).

	Male			Female		
	R	~R	rec.-rate	R	~R	rec.-rate
Treatment	48	152	(=24%)	321	188	(≈63%)
Placebo	73	145	(≈33%)	79	31	(≈72%)

Fig. 6.4 Negative effect of (the same) treatment

Put in different terms: The recovery rate is higher among those receiving the actual drug than among those receiving a placebo when N is considered in total. However, the recovery rate is higher among those receiving placebo when we look at N partitioned by gender.

The important thing to notice, of course, is that you get different results depending on the level of generalisation at which you look for associations. The main issue is that this must be a case of distortion by means of numbers. It cannot be the case, since we are analysing the exactly same data, that the drug has an overall positive as well as an overall negative identical effect on the disorder. Several problems lurk here: (1) There will always be more ways of partitioning the data by adding additional variables which might turn things upside down once more. (2) How can you know which level of abstraction is the "true" one (Malinas and Bigelow 2012; van Fraassen 2008, p. 48)?

Quickly, though, matters become even more frustrating once we start imagining potential, but perfectly plausible, partitions along less tangible dimensions than gender. For instance, current psychiatric diagnostics is created to be neutral towards aetiology and the operational methodology allows that patients receive the same diagnosis even though they only share symptomatology to a quite limited extent. Consequently, it is more than plausible that patients with distinct aetiologies are grouped together at the level of diagnosis. This would constitute a "secret" partition very difficult to control for. I will return to some complications resulting from this issue towards the end of this chapter.

The above example is hypothetical and stylised. It has been constructed specifically for the purpose of demonstrating that it is mathematically possible to get conflicting (yet statistically significant) results when a partitioned level is compared to the aggregate level. This might prompt the suspicion that there might be very strict conditions for this type of problem to occur, making it more of a theoretical conversation piece and less of a practical problem. This is not the case. In the words of Rogier Kievit and colleagues: "Simpson's paradox is not a rare statistical curiosity, but a striking illustration of our inferential blind spots [...]" (2013, p. 11).

There are reasons to believe that people without specific statistical training find it difficult to understand the challenge of Simpson's paradox. This seems to be the case even when subjects are explicitly prompted to focus on the central aspects. In (2003), Fiedler et al. studied exactly this issue and concluded that "[...] there is little evidence for a mastery of Simpson's paradox that goes beyond the most primitive level of undifferentiated guessing". Even though other studies have reached more optimistic conclusions (e.g. Spellman et al. 2001), it is quite clear that most people have considerable difficulties with engaging in "sound trivariate reasoning" (that is: analysing problems that involve the joint impact of two input variables on an output variable without committing the common fallacy of ignoring one of the input variables) (Kievit et al. 2013). On this background, at the very least, it seems reasonable to worry that non-expert users may not notice inbuilt distortions of statistical tools.

Distortion of Variance

This section will provide another example from statistics which shows that not only do certain arrangements of numbers contain tacit distortions, so do techniques for handling numbers.

Statistics is a class of tools, which are prime candidates for transferral between approaches. Numerous examples could be provided of statistical techniques originally developed for handling specific problems in a particular context, only later to spread out to be used across the disciplines. Think of correlation analysis, developed to measure the deviation of characteristics in offspring of farm animals, or factor analysis designed for use with Spearman's two-factor theory of intelligence. Both of these are now used across the board with little regard to their original application.

It is indeed a common assumption that statistical techniques are "neutral" and, thus, readily transferable between disciplines (Gigerenzer 2004).

Statistics may be used for deliberate manipulation, but the math itself is unbiased, many seem to think. I will argue to the contrary, that statistical techniques are clearcut examples of perspectival scientific tools. Indeed, they fit Giere's characterisation of what it means to incorporate a perspective more or less perfectly. As Gerd Gigerenzer has stated:

> [...] factor analysis and multiple linear regression can only "see" the linear relationships in the world, and the correlation coefficient is like an observer that ignores differences in means and variances in the two variables correlated. Thus, these methods are strictly limited tools—they let us clearly see some aspects of the world, and are mute on others. (Mitchell et al. 1997, p. 142 f.)

Considered in this way, *statistical techniques process inputs in ways peculiar to their conceptual make up, ways that render these inputs similar or different not just according to features of the inputs themselves, but also according to features of the statistical technique in question*, to paraphrase Giere once more. Statistical tools are perspectival, in other words.

As one example in prevalent use, ANOVA, short for Analysis Of Variance, is a statistical technique originally developed by Sir Ronald A. Fisher.[12] ANOVA was first developed and used to examine the manurial response of twelve different potato varieties (Fisher and Mackenzie 1923). From studies of manure, the use of ANOVA spread out through biology and the social sciences to become especially popular in psychology. A survey of the use of statistical techniques in journals published by the American Psychological Association showed that the ANOVA was used in as much as 71% of the papers by 1972 (Edgington 1974, p. 25; Haase and Ellis 1987).

Very briefly, the ANOVA is a tool for distinguishing the relative contribution of the variance of a number of causal factors to the variance of some phenotype. The ANOVA can thus (apparently) illuminate, for instance, "what proportion of the deviation of height from the population mean can be ascribed to deviation of environment from the average environment and how much to the deviation of this genetic value from the mean genetic value" (Lewontin 2006, p. 521).

One basic and central question in analyses of causation involving several contributing causes is, of course, *how* the involved variables combine to produce the phenotype. In common use of ANOVA when studying humans this question is sidestepped, however. Due to methodological shortcomings, it is widely assumed that *additivity*[13] captures (most of) the nature of causal combination. One of the attractions of additivity is that it allows the calculation of *hereditary coefficients* (equal to the variance of heredity divided by the variance of the phenotype). The hereditary coefficient is the figure presented in familiar claims such as "IQ has a 60 percent heritability", as asserted by the authors of the infamous *The Bell Curve* (Herrnstein and Murray 1994; Kievit et al. 2013, p. 5).[14]

If the relation is indeed additive one should be able to predict the effect of a genetic difference without knowing the environment of the organism in question. If, however, environment and gene interact in ways so that the environment affects gene behaviour (or vice versa) the case gets much more complicated (Griffiths and Stoltz 2013, p. 184).

Think of a gene which, if the individual is exposed to a sufficient amount of Irish folk tunes during the first years of upbringing, is expressed as a particular annoying habit of always humming and tapping the table when contemplating. If two individuals are raised in environments below that critical level, there will be no phenotypic difference between them, even though one of them carries the gene. If, on the other hand, they are brought up by barn dance enthusiasts, one will show the effect.[15]

It is, thus, by no means evident that simple additivity is the right way to construe causal combinations regarding heredity. For that matter, it is far from evident that there is one universal right way to construe causal combinations in cases of heredity.

In order to get a firm grip on the contributions of genotype and environment respectively as well as potential interactions between them, it is most often required to be able to manipulate both factors in experimentally controlled settings. When one cannot study one's phenomenon of interest in controlled settings, one must rely on observation. But when relying on observation, it is much more difficult to detect interaction. Since manipulation of nature and nurture is not normally accepted in studies of humans,[16] some statistical ingenuity is required when studying human behaviour, for instance.

The standard approach is to set up a *null hypothesis* which is basically the assumption that no relevant statistically significant variation can be detected in the data. Rejecting the null hypothesis would provide indirect support for an alternative hypothesis: that there are indeed indications of some (more or less) specified interaction in the data. In case of failure to reject the null hypothesis, however, one might conclude the absence of interaction and proceed to focus on the main effect. One may certainly question the use of null hypotheses in the first place, but that is not necessary for the point I wish to make in the following.

When dealing with null hypotheses, there are a number of standard errors to avoid. A *type I error* is to reject a null hypothesis when it is true. The significance level (α) is equivalent to what one considers to be the acceptable probability of committing a type I error. Another mistake is to accept the null hypothesis when it is false. This is a *type II error*, the probability of which is called β.

The weight of a statistically significant result depends on the statistical power of the test carried out. Statistical power is the probability of rejecting the null hypothesis when it is false (power thus equals $1 - \beta$).

Calculating statistical power is quite complex, and the calculations are not neatly written out. For the present discussion, however, it is sufficient to know that power depends on the significance level (α), the size of one's sample (N), and the population effect size (ES). The lower the α, and the higher the N and the ES, the higher the statistical power. One frequent use of power analysis, then, is in the determination of the N required to attain significant results given a hypothesised ES.

Unfortunately, when calculating statistical power in relation to using ANOVA, it is often overlooked that the power of interactions is magnitudes smaller than the power of main effects—since the ES of interactions are much smaller than the ES of main effects. The consequence is that power is often calculated solely on the basis of main effects, which leads to tests of interaction being significantly underpowered. This means not only that one cannot conclude that there is no interaction simply because one's test hereof does not show significant results to the contrary. Another consequence is, and this is much worse, that we have little reason to trust the results of studies of heredity relying on the ANOVA. Indeed, when interaction is evaluated in studies of human characteristics (such as IQ) usually no interaction is found. But laboratory studies based on controlled experimental settings consistently produce convincing evidence of various kinds of interesting interactions between nature and nurture (Wahlsten 1990, p. 112). There are few if any reasons to assume that interactions between genes and environment are less likely in humans than in, say, potatoes or rats.

The mistake is, thus, to assume that power is the same for main effect and interaction. Power is calculated on the basis of hypothesised main effect. But even in cases where power is 0.99 for the main effect, it can easily be as low as 0.40 for interaction. This means that there is a high risk of committing type II error, even though the risk appears to be negligible. In other words, there is a high risk of not rejecting the null hypothesis when it is indeed false (Wahlsten 1990, p. 116).

Since the likelihood of detecting interactions is low, interactions are rarely found. This has led to the (wrong) conclusion that a "no interaction-approach" is reasonable.

The use of null hypotheses to detect interaction is certainly a clever construction. And the ambition of detecting interaction is praiseworthy. But dealing with such advanced tools requires careful conduct. Otherwise one risks using tools in ways for which they are not suited. In the paper in which he originally presented the concept of "variance", Fisher did explicitly warn that "loose phrases about the "percentage of causation" which obscure the essential distinction between the individual and population should be carefully avoided" (Fisher 1918, p. 399 f.).

At times, the motivations behind the use of sloppy power tests is questioned:

> Transformation solely to eliminate interaction is a device to create the appearance of simplicity in the data, and there is a danger that this will be an entirely false appearance. For those who wish to learn how development actually works, wholesale and ad hoc testing of various transformations for the express purpose of getting rid of H X E interaction is counterproductive, because the shape of a functional relationship between variables provides a valuable clue to their causal connections. On the other hand, those whose only goal is to parcel out the variance among separate causes can proceed only in the absence of H X E interaction and therefore they may be more willing to transform the scale of measurement, even if causal relations become distorted. (Wahlsten 1990, p. 118)

Regardless of whether the use of ANOVA is at times deliberately fraudulent, inadequate administrations of power analysis in ANOVA tests have led prominent statisticians to state things like "[…] most of the literature on heritability in species that cannot be experimentally manipulated, for example, in mating, should be ignored" (Kempthorne 1990, p. 139), and: "The simple analysis of variance […] has no use at all. In view of the terrible mischief that has been done by confusing the spatiotemporally local analysis of variance with the global analysis of causes, I suggest that we stop the endless search for better methods of estimating useless quantities. There are plenty of real problems" (Lewontin 2006, p. 525).

The relevance of this example is multifaceted.

- First, the example shows that data treated by means of ANOVA are distorted in the sense that they appear to be additive even when they are not. This challenge is made worse by the problem that standard methods for detecting interaction fail to do so in most cases.
- Second, it is an example of a tool developed for a specific purpose which is subsequently imported into different contexts in which conditions are significantly different (e.g. in terms of the possibility of manipulation) and the potential to attain useful results is considerably lower.
- Third, since power analysis is a later development than ANOVA itself, we also have an example showing the problem that importing a tool at a given point in time does not include "automatic updating" (as discussed in Chap. 2). Though developed in the early 1960s, power analysis did not become a topic in its own right in statistics textbooks until around 1990 (Cohen 1992, p. 100). But psychologists learn their statistics mainly from psychology textbooks in which discussions of power remain rare even today (Gigerenzer 2004). Consequently, psychology students will only learn the proper careful conduct if they study beyond their curricula.
- Fourth, the example shows that tools are idealised to some extent before being applied in the new context. In this case, ANOVA is stripped of the demand of the possibility of experimentally manipulating the input parameters. Removing requirements and background assumptions means that insufficient attention is payed to important

qualifications when the idealised element is put to use in a new context.

The point I am trying to make is not that statistics is an evil. Statistical tools are certainly useful in the appropriate circumstances. The point is, that tools are perspectival in the sense of emphasising certain aspects and only being able to "see" certain kinds of relations. In the case of ANOVA as used in studies of human heredity, all causal connections are represented as additive, when additivity is probably rarely how causal factors combine.

Representing some connection as additive is perspectival in itself. And perspectives turn into distortions whenever they do not fit their targeted phenomena perfectly and whenever they leave out relevant aspects. To repeat: *statistical techniques process inputs in ways peculiar to their conceptual make up, ways that render these inputs similar or different not just according to features of the inputs themselves, but also according to features of the statistical technique in question.*

This, I believe, can be generalised as *conceptual tools process inputs in ways peculiar to their conceptual make up, ways that render these inputs similar or different not just according to features of the inputs themselves, but also according to features of the conceptual tool in question.* Conceptual tools are perspectival, in other words.

In the next section, I will present an account of an example of the relocation of a non-mathematical tool. The discussion involved is somewhat wordy, but the benefit is that we end up with corroboration for the generalisation stated just above.

The Case of Operational Definition

To further illustrate the concerns related to the transfer of tools from one context to another, I will discuss in some detail the introduction of operational definitions in the official diagnostic manuals of psychiatry since the DSM-III and the ICD-8 (Berrios and Porter 1995; Sato and Berrios 2001). The import of operational definition into psychopathology could be considered a clear example of a conceptual tool imported into a not very closely related science. This example is especially good, I believe, partly since it is relevant to the case study below, and partly because it is quite explicit what is going on.

Operational definition is also an example of what I call a propositional algorithm. You input some data into the algorithm, go through a series of specified steps, and get an output which might be processed further through other tools or fit into some representational vehicle. Further, the example of operational definitions in psychopathology makes very clear one of the dangers involved in importing propositional algorithms from quite distant approaches. Somehow, even though they were warned by Carl Hempel, the psychopathologists developing the DSM-III and subsequent versions failed to take into account the nature of inputs suitable for the algorithm.

Definitions are conceptual tools used for making clear what one is referring to in a given context—whether or not some phenomenon belongs to a category of interest. Definitions will, as a consequence, often play a significant role in representational activities. There are various types of definition, which are used more or less explicitly in different scientific settings (e.g. operational definitions, stipulative definitions, explicative definitions, and lexical definitions). As discussed in Chap. 1, in some scientific contexts you might not need definitions at all, while in other contexts they are crucial. I suggest that we think of definitions as essentially algorithmic structures, that is, as sets of rules to be followed in conceptual procedures.

In the following discussion of operational definitions, it is crucial to keep in mind that what I am referring to by 'propositional algorithm' is the algorithmic structure of a *type* of definition. I argue that the operational type of definition is ill-suited to perform the task it is used for in psychopathology. Notice also that operational definition considered as a propositional algorithm in itself does not serve to represent anything. It is a tool deliberately constructed to point out which phenomena to include in a category once a token of operational definition has been specified. But what has been transferred into psychopathology is the type of definition, not specified tokens of operational definitions once used elsewhere.

"Operational definition" is a notion developed by the physicist and 1946 Nobel laureate Percy W. Bridgman who introduced the idea in his book *The Logic of Modern Physics* (1927). Bridgman was motivated in part by the general commotion caused by the theories of relativity, as well as by specific challenges he faced as part of his work within high-pressure physics. One challenge was how to measure pressure, when one's measuring devices break as a direct result of the high pressures achieved. The central, and radical, thought Bridgman put forward was that any meaningful (theoretical) concept should be defined by a set of operations. The outcome of carrying out this set of operations would decide whether the concept applied in the specific situation or not. Bridgman believed (at first) that this was a fruitful method for replacing vague and intangible concepts.

Bridgman's formulation of the principle was the following:

> In general, we mean by any concept nothing more than a set of operations; *the concept is synonymous with the corresponding set of operations.* (Bridgman 1927, p. 5, italics in the original)

This is a quite peculiar and radical formulation, of course, and it has wide-ranging implications. The key to grasping this peculiar claim is to focus on symmetry. If the concept applies in a given situation, carrying out the set of operations would lead to a specific result. Vice versa, if the operations lead to a specific result in a given situation, the concept applies.

The algorithmic structure of operational definitions might be spelled out as such:

> ***Operational definition***: Stipulate a list of operations $\{O_1, ..., O_n\}$ as inclusion criteria for the category X and specify a (set of) result(s) $\{R\}$ which carrying out $\{O_1, ..., O_n\}$ may or may not lead to. Include any object in X if carrying out $\{O_1, ..., O_n\}$ on it, leads to (an instance of) $\{R\}$. $\{O_1, ..., O_n\}$ and their specified relations to $\{R\}$ constitutes the operational definition of X.

One might compare this to the algorithmic structure of other types of definitions:

> ***Lexical definition***: Carefully select a set of criteria $\{C_1, ..., C_n\}$ that approximates as closely as possible what is common to all the entities normally

included in the category referred to by Y in everyday language L. Use $\{C_1, ..., C_n\}$ to make distinctions between what is and what is not correctly referred to by Y in L. $\{C_1, ..., C_n\}$ constitutes the lexical definition of Y.

or

Stipulative definition: Decide on criteria $\{C_1^*, ..., C_n^*\}$ for membership of a given category Z (without worrying too much about how the term is usually used). Use $\{C_1^*, ..., C_n^*\}$ to make distinctions between what is and what is not correctly included in Z. $\{C_1^*, ..., C_n^*\}$ constitutes the stipulative definition of Z.

These types of definition are all useful, though in different circumstances. Lexical definitions, for instance, are only useful for defining terms that are already in (more or less common) use, whereas stipulative definitions can be used to define new terms (or redefine old ones). Operational definitions are certainly useful in some contexts. But when put to use in other contexts, operational definitions may have some rather regrettable consequences.

Similarly, one might spell out algorithms for different ways of observing, doing experiments, modelling data, analysing, manipulating statistics, conceptualising, constructing models, and so on . As mentioned above, some of the involved steps may involve skills, such as those for constructing and using technological equipment in physics (Collins 1985; Galison 1997). Similar arguments can be made for other sciences, for instance in psychopathological research skills regarding establishing rapport with patients (or persuading them to participate at all) are extremely important parts of establishing targets for representation in the first place. But even though there might be some parts of such processes which are skill-like in a non-propositional way and therefore cannot be easily spelled out or communicated in writing, a very large and significant part can (with some effort) be elucidated, I believe. If this were not the case, the efforts put into publication of scientific results would be in vain in quite a number of cases.

Now, Bridgman did not develop operational definition from scratch. Actually, Bridgman's achievement was more or less to generalise some fundamental ideas from Einstein's work on the special theory of relativity. Einstein's redefinitions of "length" and "time" are perhaps the most well-known examples which could be considered instances of operational definition. Indeed, Bridgman drew to a very large extent on Einstein in his book from 1927.

Operational Definition Makes Its Way into Psychopathology

As a response to a series of unpleasant challenges (e.g. Cooper et al. 1972; Rosenhan 1973), combined with a wish to rid psychiatry of the heavy influence from Freudian psychoanalysis, the American Psychiatric Association (APA) chose to base their diagnostic systems on operational definitions in the 1970s onwards. Importantly, the operational definitions should be based on "theory-free" concepts cleansed of etiological assumptions. Operational definitions were chosen in spite of the fact that this notion had been heavily criticised throughout its existence and incorporates a number of significant weaknesses, which were widely recognised long before operational definitions made their way into psychiatry (Bridgman 1945, 1954; Hempel 1954, 1959; Skinner 1945; van Fraassen 2008, p. 144).

The introduction of operational definitions in psychopathology is usually traced back to a paper presented by Carl Hempel at a 1959 WHO-conference on psychiatric nosology (Hempel 1959). In this paper, Hempel explicitly recommends the use of operational definition as a possible way of moving psychopathology towards scientific maturity. But Hempel's recommendation included some significant alterations compared to the way operational definition was used in physics.

First, Hempel stated that if operational definitions were to do any good in psychopathology, one would have to allow mere observations (as compared to measurements and experiments) to count as operations. Second, and very importantly, he shifted the conceptual level at which operational definitions were to be put to use. Einstein, and to a large extent Bridgman, used operational definition to define or redefine very basic, fundamental concepts by means of quite simple operations. Hempel, on the other hand, suggested defining *diagnoses* operationally. Diagnoses are concepts of a very different and less fundamental nature than, say, "time" or "length". I will return to the consequences of this shift below.

Expressed in universal terms as it is, Bridgman's statement is quite absurd. We do not need operational definitions for terms like 'hot dog' or 'Donald Trump' in order for these to be meaningful. There are, however, very good reasons to think that Bridgman did not intend this as a general semantic theory, partly because later on he explicitly denied having had such intensions (Bridgman 1954). It is probably the case, that Bridgman operated with an implicit quantifier domain restriction. That is, by "any concept" he might have meant something like "all sufficiently important and fundamental concepts in physics".

Speculative interpretations aside, Bridgman's use of the universal quantifier nevertheless impressed Hempel to the extent that he felt he had to warn against a potentially threatening regress: that the terms used to define something operationally themselves required operational definition and so on (perhaps ad infinitum).

In his 1959 paper, Hempel pointed explicitly to the problem of the regress. But Hempel also pointed out a potential solution of basing operational definitions on a foundation of concepts, which require no further definition. Hempel perspicuously emphasised that not all concepts will serve equally well as the regress-stopping "certain" foundation on which operational definitions may be based. Hempel wrote:

> In any definitional context (quite independently of the issue of operationism), some terms must be taken to be antecedently understood; and the objectivity of science demands that the terms which thus serve as a basis for the introduction of other scientific terms be used with a high degree of uniformity by different investigators in the field. (Hempel 1959, p. 11)

So far, so good. Hempel is right in diagnosing the problem of the regress (obviously), and also right in prescribing a treatment of ultimately basing one's definitions on concepts which might be defined in non-operational manners. But, as we shall see below, that terms are used with a high degree of uniformity do not guarantee their capacities as the basis for distinguishing accurately between complex phenomena.

Psychiatrist Erwin Stengel arranged the 1959 conference and summarised the results in a rapport the same year. His conclusions were very true to Hempel's suggestions. But with this report he provided a crucial link by which operational definition could enter further into psychopathology.

Another psychiatrist, Sir Aubrey Lewis who also attended the 1959 conference, was subsequently involved in a WHO programme on psychiatric epidemiology. During this work, he proposed a dual diagnostic system where global cross cultural issues were dealt with by means of an operationalised a-theoretical part based on non-specialist terms for ease of use, while research was to be carried out using a theory-based version.

WHO and APA followed Lewis' recommendations, but discarded the research-version. They opted for a single system incorporating operationally defined diagnoses based on observational/descriptive theory-neutral layman terms. This system is, well, operational today (Fulford and Sartorius 2009).

Obviously, even though a few additional steps are added to the story as it is normally told, a lot of detail is still omitted. Nevertheless, an interesting picture emerges if we look closer at how the use of the operational definition algorithm changed at each of these steps.

Let me, therefore, try to sum this up. I claim that the operational definition algorithm is virtually unchanged from its origin in physics to its contemporary use in psychopathology. In the discussion above we have traced operational definitions back to Einstein's Special Theory of Relativity. Now, Einstein's use of operationalisation was completely legit and healthy in every sense. So, if the algorithm is itself well-shaped, what else is out of order and can account for the claim that operational definition is dysfunctional in psychopathological diagnostics?

A close look at how *the use* of operational definitions has changed will be revealing.

First we have Einstein with the fruitful assumption that we might gain something by operationally redefining certain central terms.

Bridgman then made the sweeping (unintentional) generalisation that *all* terms must be operationally defined in order to be meaningful.

As an antidote, Hempel recommended basing operational definitions on antecedently understood terms which are used with a high degree of uniformity among "investigators in the field". As mentioned, acknowledging the limited possibilities for direct measurements and experiments in psychopathology, Hempel added the point about allowing observations to count as input into operational definitions in the context of psychopathology.

Sir Aubrey Lewis added the idea of basing operational definitions on non-specialist theory neutral concepts for general statistical purposes, with another etiologically informed version for research purposes. The theory-neutral half was eventually chosen as the exclusive basis for ICD-8 and DSM III as well as the subsequent editions of psychiatric diagnostic manuals.

If we compare the first and the last steps in this development, it is evident that the use of operational definition is significantly different in contemporary psychopathology compared to its original context. There is, of course, no reason to assume that a specific way of using a tool is the best one, simply because it was first. But in this case, there can be little doubt that the theories of relativity are superior to DSM 5 (to phrase it in a somewhat popular way).

In the original version, the algorithm is fed simple, publicly observable actions and measurements in order to define primitive, fundamental concepts. In psychopathology, the algorithm is fed vague, non-specialist layman terms denoting subjective observation in order to define complex and poorly understood symptom clusters.

In his (1959) paper, Carl Hempel used the following example to illustrate the nature of operational definitions:

> A simple operational definition of the term *harder than* as used in mineralogy might specify that a piece of mineral x is to be called harder than another piece of mineral y, if the operation of drawing a sharp point of x under pressure across a smooth surface of y has as its outcome a scratch on y, whereas y does not thus scratch x. (p. 8)

This is a very nice example, since it exemplifies the advantages as well as the limitations of operational definitions. In mineralogy, this method for deciding which of two minerals is the hardest has proven its worth long ago. Indeed, the method dates back at least to year 300 BC where

Theophrastus (who succeeded Aristotle as head of the academy in Athens) mentioned the method in his treatise *On Stones*. Reputedly, this method works quite well for identifying minerals in the field even today.

Notice, however, how the recipe for deciding which of two minerals are the hardest presuppose that we already know (or at least agree) that the two objects compared are both instances of the category of minerals. Notice also, that the method for determining which of the two objects is the hardest relies heavily on the result of an experiment, which is directly publicly observable by anyone with a well-functioning sense-apparatus.

None of these characteristics apply to psychopathology. For instance, in spite of occasional disagreement over taxonomic issues in mineralogy (Nickel 1995a, b), there are no anti-mineralogy-movements and no deniers of the existence of minerals as such. Further, the operational definitions in psychopathology are not just supposed to determine which of a number of possible disorders a person suffers from, but also whether he or she suffers from anything at all.

Now, compared to the hardness of minerals, mental disorders or pathologies are much more complex phenomena tangled up in innumerable interacting causal processes. This is no reason to believe that, literally or metaphorically, drawing a sharp point of one psychiatric patient under pressure across a smooth surface of another psychiatric patient would provide us with any interesting information about their respective pathologies.

Stupidities aside, Hempel's example quite clearly shows the utility of operationalisation in one specific setting. Ironically, the example also demonstrates why operational definition cannot be expected to be useful for the solving of psychopathological taxonomic difficulties. It ought not be necessary to make silly jokes in order to make plain the limits of Hempel's analogy in relation to psychopathology.

Current Problems Facing Operational Definition in Psychopathology

If a set of operations defines a concept, then carrying out the set of operations and registering the result will tell us whether the concept applies or not. On the one hand, this corresponds neatly to filling out an ADHD-questionnaire and deciding on a diagnosis from looking at the pencil marks on the paper. The central propositional algorithm of Operational Definition is intact in its contemporary use in psychiatry.

On the other hand, of course, there does seem to be somewhat of a tension here. People tend to construe being diagnosed with ADHD as something more than this. You rarely hear anybody say "I have ADHD. ADHD is defined as a quite heterogeneous class of combinations of answers to a certain questionnaire". This is the consequence of operational definition being a very artificial type of definition, quite far from how the human conceptual system naturally operates. It does not mean that it is useless in scientific settings, of course.

But any algorithm is heavily dependent on the inputs it is fed. In contemporary diagnostics, the inputs into the operational definition algorithm consist to a very large extent of everyday concepts such as "nervous", "indecisive", and "inattentive". The reasoning behind the choice of lay terminology is based on the assumption that since everyday concepts are concepts we are all intimately familiar with, they can be expected to be applied stably by all subjects. Everybody knows what it means to be nervous. This is a grave mistake—a mistake which can be traced directly back to Hempel's recommendation of "antecedently understood terms that are used with a high degree of uniformity among investigators in the field". In contemporary psychopathology, the "investigators in the field" include the parents, teachers, relatives, and so on, who are asked to fill out the questionnaires which are the central tools in parts of contemporary diagnostics. These people are indeed, with all respect, genuine laymen. And the terminology that is used with a high degree of uniformity in this group of people is, of course, lay terminology.

Lay expressions such as 'often interrupts' or 'often forgetful' or 'often fails to give close attention to details' covers very wide sets of phenomena. Some of the phenomena included are completely healthy, others are indeed more or less pathological. There is a huge difference between the reliability, stability, and conceptual coordination of basic measurements of contemporary physics and the concepts that serve as the inputs into the operational definitions in psychopathology. Remember, it is not the case, that input terms in psychopathology are used in special "specialist's senses" which simply recycle lay terminology. Indeed, as just mentioned, it is to a large extent genuine laymen who carry out diagnostics in contemporary psychiatry by assessing various difficulties via multiple-choice tests.

Even though the algorithm of operational definition is a useful idea in the right contexts, it is no more reliable than the inputs it is fed. And there is no feature in the algorithm for distinguishing between manifestations of pathological issues and perfectly healthy and normal responses to challenges of everyday life.

Final Remarks on OD?

This, then, has been a story about relocating a conceptual tool, operational definition, from one scientific context to another. The central motivation for this account, of course, is that it substantiates the claim that conceptual tools are perspectival. The perspective of operational definitions is well suited to define or redefine central basic concepts when fed accurate and reliable inputs. On the other hand, it is not suited for defining complex concepts by being fed vague and indeterminate inputs.

Part of the explanation for the operational degeneration described is probably a lack of sensitivity to the special requirements of the different contexts in which operational definition is put to work. Though the central algorithm remains unchanged through the process, the inputs fed into the algorithm are of very different kinds in the different contexts. Further, the conceptual level at which the algorithm is applied is also very different from the original use. Indeed, Hempel knew very little about psychiatry, as he explicitly stated in his nosology paper. The people eventually responsible for deciding on putting operational definition to work in ICD 8 & DSM III have probably not been aware of the important differences between defining fundamental, primitive concepts by means of operational definition, and defining higher level complex concepts by the same method.

It is my claim, that the present use of operational definitions in psychopathology bears a significant part of the responsibility for the conceptual inaccuracy which plagues contemporary psychiatry and psychopathology.

And even though this is not an exhaustive explanation for the dramatic increase in diagnoses such as depression and ADHD, the use of operational definitions based on lay terminology to define complex concepts certainly facilitates diagnostic explosions due to its inherent over-inclusive nature (Haslam 2013; Parnas and Bovet 2014; Sato and Berrios 2001).

The above should give an impression of what I refer to by 'propositional algorithm'. Propositional algorithms constitute one of the types of element I claim make up the intermediate layer of scientific representation. By analogy, one might think of the involved elements as having specific functions like the different parts of a car engine. There are different types of car engines that require different types of parts to perform their overall function, that is, to convert energy into motion. Some of these parts are interchangeable, while others are important in one type of engine and superfluous in others (e.g. spark-plugs without which a gasoline engine will not work, but which serve no purpose in a diesel engine). Similarly, some elements might be indispensible in some scientific approaches, but superfluous in others; transferable between some scientific approaches, but not between others.

As mentioned above, it is hard to see principled limitations as to which kinds of elements might be transferred and from where to where. *Insight A* and *Insight B* tells us that we cannot easily conclude how well elements will perform in new contexts. We need to determine in detail *how* elements are used, which purposes they serve, and how they are combined. Determining this, will bring one significantly closer to being able to assess how well-formed a given interdisciplinary approach is.

Simpson's Revisited

My discussion of operational definitions has been quite lengthy. But I am not letting the patient reader off the hook just yet. Before letting the topic of operational definition rest, I want to draw out a further consequence related to concealed distortions which may, in some cases, be a consequence of the relocation of theoretical elements. This consequence of using operational definitions in psychiatry is relevant to the discussions of Simpson's paradox and ANOVA above.

As is well known, there is quite a bit of public debate regarding whether to medicate people diagnosed with mental illnesses. The opposition to medication in psychiatry is represented most saliently by anti-psychiatrists such as Thomas Szasz (Szasz 1960, 1961), who is infamous for outright denying that mental illnesses are real phenomena. But people with significantly less radical convictions also raise doubts about whether to medicate, for instance, people diagnosed with depression or ADHD.

In Denmark, Peter Gøtzsche recently released a controversial book (2015) in which he concludes that there is no evidence that medication prescribed for depression has any beneficial effects for the patients. He bases his conclusions on a metastudy of a large number of published and unpublished studies of double-blinded trials of the effects of antidepressants.

In a double-blinded study neither the participants nor the researchers studying them are supposed to know who gets the placebo and who gets the active ingredients. But as it turns out, the active ingredients of antidepressants have a number of easily recognisable side-effects, which in many cases make it easy for the researcher to figure out which kind of pill a participant has received.

Gøtzsche claims, that once corrected for biases of such pseudo-double-blinded approaches, the positive effects of antidepressants vanish completely. All that remain are the negative side-effects. As a consequence, Gøtzsche concludes, all treatments with antidepressants should be stopped.

Interestingly, however, since diagnostics are based on operational definitions which are (1) based on very broad input concepts, and (2) deliberately cut off from aetiology, we end up in a situation where Gøtzsche's conclusion cannot be drawn conclusively. And what is worse, the operational non-etiological approach prevents us from concluding anything.

As a consequence of (1), it is a quite heterogenous group of people who is diagnosed as suffering from depression. As a consequence of (2), it is unreasonably difficult to figure out to what extent the causal pathways leading the patients to be diagnosed are similar. The only formally accepted method for deciding whether to include someone in a sample of people suffering from depression is the official operationalised taxonomic systems.

Discussions of differences in aetiology are rarely included in studies of the effects of antidepressants. Rather, "major depressive disorder" is considered more or less as a natural kind for which we are searching for the one effective treatment.

It is thus misleading to say that depression may manifest itself in various ways. It would be far more correct to state that a large number of divergent phenomena may lead someone to being diagnosed with major depressive disorder *and* we do not know the extent to which these expressions of distress are related at the causal level. Therefore, we do not know, either, to what extent some medication might have different effects depending on distinct aetiology. In other words, while we have means to decide which of a group of applicants are men and which are women, the contemporary taxonomic system provides no means to decide whether a cohort of people diagnosed with major depressive disorder do so as an effect of cause A, B, or C, and so on.

Say, for the sake of simplicity (although we have little reason to assume that reality is this simple), that there are two distinct causal pathways, CP_a and CP_b, which both lead to a diagnosis of major depressive disorder. Since we do not distinguish between CP_a and CP_b in clinical trials of antidepressants, it might be the case that the drug has a beneficial effect on CP_a-patients and a negative on CP_b-patients which would add up to the absence of positive effects that Gøtzsche reports. But, and this a substantial complication, as we have learned from the discussion of Simpson's paradox above, it is even possible to have a positive effect in both subgroups while seeing negative effect (or no effect) in the aggregated group.

Due to the nature of psychiatric diagnostics, then, we cannot decide whether Gøtzsche's conclusions are correct. There is little doubt that he points out important and severe difficulties inherent to psychiatric research. But his conclusions regarding the effects of anti-depressants may be wrong due to tacit diagnostic distortions, and we can only escape this predicament if we reconstruct the entire psychopathological system in more exact and, importantly, aetiology-based terms—and, of course, do all the existing studies all over again. One might wonder whether difficulties like these contribute to the current replication crisis of the pharmaceutical industry (Begley and Ellis 2012; Osherovich 2011; Prinz et al. 2011).

Summing Up

Since all acts of representation involve (more or less unapparent) distortions, one could provide innumerable further examples. For now, let the above suffice to show that for the non-expert it is sometimes far from evident which unapparent distortions are imported along with perfectly proportional surface transformations.

That some salient surface characteristics, like geometric proportions, are invariant through a given representational transformation might divert attention from the fact that less apparent properties, like strength, are changed through the process. As another example: the surface characteristic that the same statistical technique is used to analyse two different data sets may divert attention from the fact that the possibility of experimental manipulation makes a significant difference regarding what can be concluded about causal connections in the two cases.

When we add to this that interdisciplinarity most often involves dealing with theoretical elements from outside one's central field of expertise, we may safely assume that the appropriate insights are not always available. This is equally problematic for the person pushing her expertise onto a field different from her home discipline, as for the person importing theory of which she has only limited understanding.

One might also imagine that there are degrees of problems with relocation. It is probably much more challenging to relocate propositional structure from quantum mechanics to musicology than from say, physics to biology or from biology to economics. Due to the larger distance between the parent approaches, much more of the tacit knowledge of quantum mechanics is likely to be lost in the process. This is likely to result in the use of superficially understood theoretical elements.

There are, thus, very good reasons for increasing the focus on representation, and the issue of distortion, in connection with analyses of interdisciplinarity: In the combination of two or more approaches, all aspects of the involved acts of representation are in play. This includes idealisations, approximations, distortions, vehicles, tools, algorithms, and basic assumptions. One cannot simply identify and replace distorted elements of one approach with "undistorted" elements of another. Further, two combined distortions cannot be assumed to level each other out.

Presumably, some sort of interference will occur between the (distortions of the) combined elements when approaches are integrated. Whether this will result in more or less valid, robust, or relevantly purpose-serving approaches must be analysed carefully on a case-by-case basis. A good place to start such a case-by-case analyses would be by identifying the vehicles and intermediate elements of the integrated approach, and figuring out how they are used.

Approach-based analysis thus identifies a number of aspects to focus on in analysis of interdisciplinary activities. This brings us considerably closer to an adequate method for assessing epistemic aspects of interdisciplinarity. It also reveals a number of pressing difficulties which ought to generate ample motivation for caution in the development, evaluation, and execution of interdisciplinary research projects.

Notes

1. ... such as "quarks" (Pickering 1984), "gender" (Lorber and Farrell 1991), "scientific knowledge" (MacKenzie 1981), and, of course, "reality" (Berger and Luckmann 1966). In most cases, though, the positions seem to be that certain *concepts* or *conceptual frameworks* are socially constructed, rather than the entities in question themselves. The most level-headed discussions of these issues are (Hacking 1999) and (Collin 1997), I believe.

2. On separate occasions, at seminars I have attended, I have heard as well Bruno Latour as Harry Collins assuring that (of course) they are (ontological) realists, and that anything else would be foolish.

3. Gilbert and Mulkay focus on developments and disagreements within research on oxidative phosphorylation of all things.

4. For instance, monochromatic light with a wavelength of 590 nm is experienced as orange. So is a combination of the wavelengths 670 nm (red) and 546 nm (green). At certain red/green-ratios, people with normal trichromatic vision have trouble discriminating between these. This is the phenomena known as metamerism (Giere 2006b, p. 21 f.). A study of anomalous human vision showed that at least one person discriminates very fast and reliably between orange light with different spectral characteristics. This at least indicates that human (behavioural) tetrachromacy does exist (Jordan et al. 2010).

5. If this is not obvious, I recommend Giere's very careful discussion hereof in (Giere 2006b) as well as, perhaps, *The Island of the Colorblind* by Oliver Sacks (1997).

6. http://thetruesize.com/ is a fun web-resource for experimenting with the distortions of map projections.

7. I realise that it might be somewhat tiring that I keep returning to this equation. However, Giere (and others as well) has spent a few decades analysing this example for which reason I cannot simply ignore it here.

8. In which some properties or elements are omitted from representation leaving only the "essential" ones. This type of idealisation is sometimes referred to as 'abstraction' (Cartwright 1989; Jones 2005). I, however, prefer to use 'idealisation' for several reasons. (1) I agree with Weisberg's views about the utility of a pluralistic account of idealisation (Weisberg 2007), (2) I think 'abstraction' is a potentially confusing term, given the many senses with which it is related, and (3) I believe it makes best sense to use 'idealisation' to describe all the different ways that lead to the construction of "ideal systems" (which are also the results of abstraction or omission, as I see it).

9. A good source for these kinds of considerations is (Jones 2005).

10. Diathesis-stress-models come in many more or less specified versions and goes back at least to Pierre Briquet's systematic studies of hysteria in 1859 (Ellenberger 1970, p. 142). More recently Paul Meehl's discussion of Schizophrenia is a well-known example (Meehl 1962). Presently, I make use of a strongly idealised version of the diathesis-stress-model (allegedly capturing general aspects of the relationship between vulnerability and breakdown) since it provides the best basis for comparison with the equally strongly idealised model of the simple pendulum.

11. The figures in this example are borrowed from (Malinas and Bigelow 2012).

12. Fisher is an interesting example since he himself drew heavily on statistical thermodynamics in developing his contributions to genetics (Depew and Weber 1995; Griffiths and Stoltz 2013, p. 2)

13. Additivity means that the variance of the phenotype can be calculated simply by adding up the variance of environment and the variance of the genotype. (Griffiths and Stoltz 2013, p. 183)

14. Since what is truly interesting is the contribution of genes and environment respectively, there is a tendency to forget that the deviation of some value from the mean is not the same as the contribution of the value. The

hereditary coefficient would increase in a more uniform environment due to more of the phenotypic variance being ascribed to the genetic variance. If people, on the other hand, were more genetically similar, the hereditary coefficient would decrease due to less genetic difference to correlate with phenotypic differences. This, however, tells us little about the actual genetic contribution. For instance, increasing equality in terms of access to education increases the heritability of IQ (Griffiths and Stoltz 2013, p. 185).

15. The nasty habit will, of course, slowly but surely interact with the gene carrier's environment, which will turn increasingly hostile towards him.

16. In studies of human behaviour, ethical considerations prevent otherwise respectable methods such as experimental control of mating patterns as well as radical manipulation of nurture.

References

Begley, C.G., and L.M. Ellis. 2012. Drug Development: Raise Standards for Preclinical Cancer Research. *Nature* 483: 531–533. https://doi.org/10.1038/483531a.

Berger, Peter L., and Thomas Luckmann. 1966. *The Social Construction of Reality: A Treatise in the Sociology of Knowledge*. Garden City, NY: Anchor Books.

Berrios, G.E., and Roy Porter. 1995. *A History of Clinical Psychiatry: The Origin and History of Psychiatric Disorders*. London: Athlone.

Bridgman, P.W. 1927. *The Logic of Modern Physics*. New York: The Macmillan Company.

———. 1945. Some General Principles of Operational Analysis. *Psychological Review* 52 (5): 246–249.

———. 1954. Remarks on the Present State of Operationalism. *The Scientific Monthly* 79 (4): 224–226.

Cartwright, Nancy. 1983. *How the Laws of Physics Lie*. Oxford; New York: Clarendon Press; Oxford University Press.

———. 1989. *Nature's Capacities and their Measurement*. Oxford; New York: Clarendon Press; Oxford University Press.

———. 1999. *The Dappled World: A Study of the Boundaries of Science*. Cambridge; New York: Cambridge University Press.

Cohen, Jacob. 1992. Statistical Power Analysis. *Current Directions in Psychological Science* 1 (3): 98–101.

Collin, Finn. 1997. *Social Reality, The Problems of Philosophy*. London; New York: Routledge.

Collins, Harry M. 1985. *Changing Order: Replication and Induction in Scientific Practice*. London: The University of Chicago Press.

Contessa, Gabriele. 2006. Scientific Models, Partial Structures and the New Received View of Theories. *Studies in History and Philosophy of Science* 37 (2): 370–377.

Cooper, J.E., R.E. Kendell, B.J. Gurland, L. Sharpe, and J. Copeland. 1972. *Psychiatric Diagnosis in New York and London; A Comparative Study of Mental Hospital Admissions, Maudsley Monographs*. London; New York: Oxford University Press.

Depew, David J., and Bruce H. Weber. 1995. *Darwinism Evolving: Systems Dynamics and the Genealogy of Natural Selection*. Cambridge, MA: MIT Press.

Downes, Stephen M. 1992. The Importance of Models in Theorizing: A Deflationary Semantic View. *PSA 1992: Proceedings of the Biennial Meeting of the Philosophy of Science Association* 1: 142–153.

Edgington, E.S. 1974. A New Tabulation of Statistical Procedures in APA Journals. *American Psychologist* 29: 25–26.

Ellenberger, Henri F. 1970. *The Discovery of the Unconscious: The History and Evolution of Dynamic Psychiatry*. London: Allen Lane, The Penguin Press.

Fiedler, Klaus, Eva Walther, Peter Freytag, and Stefanie Nickel. 2003. Inductive Reasoning and Judgment Interference: Experiments on Simpson's Paradox. *Personality & Social Psychology Bulletin* 29 (1): 14–27.

Fisher, Ronald A. 1918. The Correlation between Relatives on the Supposition of Mendelian Inheritance. *Transactions of the Royal Society of Edinburgh* 52: 399–433.

Fisher, Ronald A., and W.A. Mackenzie. 1923. Studies in Crop Variation: II. The Manurial Response of Different Potato Varieties. *Journal of Agricultural Science* 13: 311–320.

Fulford, K.W.M., and Norman Sartorius. 2009. The Secret History of ICD and the Hidden Future of DSM. In *Psychiatry as Cognitive Neuroscience*, ed. Matthew R. Broome and Lisa Bortolotti. Oxford: Oxford University Press.

Galison, Peter. 1997. *Image and Logic: A Material Culture of Microphysics*. Chicago; London: University of Chicago Press.

Giere, Ronald N. 1988. *Explaining Science: A Cognitive Approach, Science and Its Conceptual Foundations.* Chicago: University of Chicago Press.

———. 1999. *Science without Laws, Science and Its Conceptual Foundations.* Chicago: University of Chicago Press.

———. 2006a. Perspectival Pluralism. In *Scientific Pluralism*, ed. Stephen H. Kellert, Helen Longino, and C. Kenneth Waters, 25–41. Minneapolis, MN: University of Minnesota Press.

———. 2006b. *Scientific Perspectivism.* Chicago: University of Chicago Press.

Gigerenzer, Gerd. 2004. Mindless Statistics. *The Journal of Socio-Economics* 33: 587–606.

Gilbert, G. Nigel, and Michael Mulkay. 1984. *Opening Pandora's Box: A Sociological Analysis of Scientists' Discourse.* Cambridge: Cambridge University Press.

Godfrey-Smith, Peter. 2006. The Strategy of Model-Based Science. *Biology and Philosophy* 21: 725–740.

Gøtzsche, Peter C. 2015. *Dødelig psykiatri og organiseret fornægtelse.* Denmark: People's Press.

Griesemer, James R. 1990. Modeling in the Museum: On the Role of Remnant Models in the Work of Joseph Grinnell. *Biology & Philosophy* 5: 3–36.

Griffiths, Paul, and Karola Stoltz. 2013. *Genetics and Philosophy: An Introduction.* Edited by Michael Ruse. Cambridge Introductions to Philosophy and Biology. Cambridge: Cambridge University Press.

Haase, Richard F., and Michael V. Ellis. 1987. Multivariate Analysis of Variance. *Journal of Counseling Psychology* 34 (4): 404–413.

Hacking, Ian. 1999. *The Social Construction of What?* Cambridge, MA: Harvard University Press.

Haslam, Nick. 2013. Reliability, Validity, and the Mixed Blessings of Operationalism. In *The Oxford Handbook of Philosophy and Psychiatry*, ed. K.W.M. Fulford, Martin Davies, Richard G.T. Gipps, George Graham, John Z. Sadler, Giovanni Stanghellini, and Tim Thornton. Oxford: Oxford University Press.

Hempel, Carl G. 1954. A Logical Appraisal of Operationism. *The Scientific Monthly* 79 (4): 215–220.

Hempel, Carl R. 1959. Introduction to Problems of Taxonomy. In *Field Studies in the Mental Disorders*, ed. J. Zubin, 3–23. New York: Grune and Stratton.

Herrnstein, Richard J., and Charles Murray. 1994. *The Bell Curve: Intelligence and Class Structure in American Life.* New York: Free Press.

Ingram, R.E., and D. Luxton. 2005. Vulnerability-Stress Models. In *Development of Psychopathology: A Vulnerability-Stress Perspective*, ed. Benjamin L. Hankin and John R. Z. Abela, x, 510 p. New York: Sage Publications.

Jones, Martin R. 2005. Idealization and Abstraction: A Framework. In *Idealization XII: Correcting the Model: Idealization and Abstraction in the Sciences*, ed. Martin R. Jones and Nancy Cartwright, 173–217. New York: Rodopi.

Jordan, Gabriele, Samir S. Deeb, Jenny M. Bosten, and J.D. Mollon. 2010. The Dimensionality of Color Vision in Carriers of Anomalous Trichromacy. *Journal of Vision* 10 (8): 1–19.

Kellert, Stephen H., Helen Longino, and C. Kenneth Waters, eds. 2006. *Scientific Pluralism, Minnesota Studies in the Philosophy of Science*. Minneapolis, MN: University of Minnesota Press.

Kempthorne, Oscar. 1990. How Does One Apply Statistical Analysis to Our Understanding of the Development of Human Relationships. *Behavioral and Brain Sciences* 13: 138–139.

Kievit, Rogier A., Willem E. Frankenhuis, Lourens J. Waldorp, and Denny Borsboom. 2013. Simpson's Paradox in Psychological Science: A Practical Guide. *Frontiers in Psychology* 4: 513.

Lenhard, Johannes, Günther Küppers, and Terry Shinn, eds. 2010. *Simulation: Pragmatic Construction of Reality*. Dordrecht, The Netherlands: Springer.

Levy, Arnon, and Adrian Currie. 2014. Model Organisms are Not (Theoretical) Models. *The British Journal for the Philosophy of Science* 0: 1–22.

Lewontin, R.C. 2006. The Analysis of Variance and the Analysis of Causes. *International Journal of Epidemiology* 35: 520–525.

Lorber, Judith, and Susan A. Farrell, eds. 1991. *The Social Construction of Gender*. Newbury Park, CA: Sage Publications.

MacKenzie, Donald. 1981. *Statistics in Britain, 1865–1930: The Social Construction of Scientific Knowledge*. Edinburgh, UK: Edinburgh University Press.

Malinas, Gary, and John Bigelow. 2012. Simpson's Paradox. In *The Stanford Encyclopedia of Philosophy* (Winter 2012 Ed.), ed. Edward N. Zalta. http://plato.stanford.edu/archives/win2012/entries/paradox-simpson/

McMullin, Ernan. 1985. Galilean Idealization. *Studies in History and Philosophy of Science* 16: 247–273.

Medin, Douglas L., Elizabeth B. Lynch, John D. Coley, and Scott Atran. 1997. Categorization and Reasoning among Tree Experts: Do All Roads Lead to Rome? *Cognitive Psychology* 32: 49–96.

Meehl, P.E. 1962. Schizotaxia, Schizotypy, Schizophrenia. *American Psychologist* 17 (12): 827–838. https://doi.org/10.1037/h0041029.

Mitchell, Sandra D. 2002. Integrative Pluralism. *Biology and Philosophy* 17 (1): 55–70.

———. 2012. *Unsimple Truths*. 1 vol. Chicago, IL; Bristol: University of Chicago Press; University Presses Marketing distributor.

Mitchell, Sandra D., Lorraine Daston, Gerd Gigerenzer, Nevin Sesardic, and Peter Sloep. 1997. The Why's and How's of Interdisciplinarity. In *Human By Nature: Between Biology and the Social Sciences*, ed. Peter Weingart, Sandra D. Mitchell, Peter J. Richerson, and Sabine Maasen, 103–150. Mahwah, NJ: Erlbaum Press.

Morgan, Mary S., and Margaret Morrison, eds. 1999. *Models as Mediators: Perspectives on Natural and Social Science. Ideas in Context*. Cambridge: Cambridge University Press.

Nickel, Ernest H. 1995a. The Definition of a Mineral. *The Canadian Mineralogist* 33: 689–690.

———. 1995b. Mineral Names Applied to Synthetic Substances. *Mineralogy and Petrology* 57: 261–262.

Osherovich, L. 2011. Hedging against Academic Risk. *Science-Business eXchange* 4 (15). https://doi.org/10.1038/scibx.2011.416.

Parnas, Josef, and Pierre Bovet. 2014. Psychiatry Made Easy: Operation(al)ism and Some of Its Consequences. In *Philosophical Issues in Psychiatry III: The Nature and Sources of Historical Change*, ed. Kenneth S. Kendler and Josef Parnas. Oxford: Oxford University Press.

Pickering, Andrew. 1984. *Constructing Quarks: A Sociological History of Particle Physics*. Chicago, IL: University of Chicago Press.

Prinz, F., T. Schlange, and K. Asadullah. 2011. Believe It or Not: How Much Can We Rely on Published Data on Potential Drug Targets? *Nature Reviews Drug Discovery* 10: 712–713.

Reutlinger, Alexander, and Matthias Unterhuber. 2014. Thinking about Non-Universal Laws: Introduction to the Special Issue Ceteris Paribus Laws Revisited. *Erkenntnis* 79: 1703–1713. https://doi.org/10.1007/s10670-014-9654-5.

Rosenhan, David L. 1973. On Being Sane in Insane Places. *Science* 179: 250–258.

Sacks, Oliver. 1997. *The Island of the Colorblind*. 1st ed. New York: A.A. Knopf: Distributed by Random House.

Sato, Y., and G.E. Berrios. 2001. Operationalism in Psychiatry: A Conceptual History of Operational Diagnostic Criteria. *Clinical Psychiatry* 43: 704–713.

Simpson, E.H. 1951. The Interpretation of Interaction in Contingency Tables. *Journal of the Royal Statistical Society Series B—Statistical Methodology* 13 (2): 238–241.

Skinner, B.F. 1945. The Operational Analysis of Psychological Terms. *Psychological Review* 52 (5): 270–277.

Spellman, B.A., C.M. Price, and J. Logan. 2001. How Two Causes are Different From One: The Use of (Un)conditional Information in Simpson's Paradox. *Memory and Cognition* 29: 193–208.

Suppe, Frederick. 1972. Whats Wrong with the Received View on the Structure of Scientific Theories? *Philosophy of Science* 39 (1): 1–19.

Szasz, Thomas. 1960. The Myth of Mental Illness. *American Psychologist* 15: 113–118.

———. 1961. *The Myth of Mental Illness: Foundations of a Theory of Personal Conduct*. New York: Dell.

Thomson-Jones, Martin. 2012. Modeling without Mathematics. *Philosophy of Science* 79: 761–772.

van Fraassen, Bas C. 2008. *Scientific Representation: Paradoxes of Perspective*. Oxford; New York: Clarendon Press; Oxford University Press.

Wahlsten, Douglas. 1990. Insensitivity of the Analysis of Variance to Heredity-Environment Interaction. *Behavioral and Brain Sciences* 13: 109–161.

Weisberg, Deena Skolnick. 2008. Caveat Lector: The Presentation of Neuroscience Information in the Popular Media. *The Scientific Review of Mental Health Practice* 6 (1): 51–56.

Weisberg, Michael. 2007. Three Kinds of Idealization. *The Journal of Philosophy* 104 (12): 639–659.

———. 2013. *Simulation and Similarity: Using Models to Understand the World. Oxford Studies in Philosophy of Science*. New York: Oxford University Press.

Wimsatt, William. 2007. *Re-engineering Philosophy for Limited Beings: Piecewise Approximations of Reality*. Cambridge, MA: Harvard University Press.

Winsberg, Eric. 2010. *Science in the Age of Computer Simulation*. Chicago, IL: The University of Chicago Press.

7

Representational Crossbreeding

This chapter is supposed to make clear how to apply approach-based analysis to concrete cases of interdisciplinary science. It is time, finally, to resurface from the depths of scientific representation. Against the background of the discussions above, we shall attempt to arrive at an overview and a concise guideline for how to evaluate epistemic aspects of interdisciplinary cases.

The framework presented in this chapter builds on (and presupposes) the more technical discussions of Chaps. 5 and 6. I recommend reading the following with the preceding discussions fresh in mind. If you have skipped the preceding chapters, I recommend giving them a second shot before reading the following.

The discussion in this chapter will provide the structure of approach-based analysis. It will be quite abstract, though. In the following chapter, the developed method will be applied in a case study of a specific interdisciplinary approach in order to demonstrate how such an analysis might be carried out in practice.

The framework presented will highlight potential pitfalls in relation to interdisciplinary work by pointing out problems related to relocation of representational elements. This may appear uninspiring and dull to those who just want to go ahead and be innovative. The aim is not to be overly

© The Author(s) 2018
R. Hvidtfeldt, *The Structure of Interdisciplinary Science*, New Directions
in the Philosophy of Science, https://doi.org/10.1007/978-3-319-90872-4_7

conservative or restrictive. But applying approach-based analysis indicates that some caution is in place. Science, including interdisciplinary science, requires meticulous care and attention. Everyone involved in scientific activities knows this, of course. What is highlighted by approach-based analysis is a number of unapparent pitfalls specific to interdisciplinary science which require more attention than they usually get. The extent to which interdisciplinary researchers actually commit such mistakes is obviously an empirical question. Locating the pitfalls, however, is a prerequisite for determining whether researchers actually fall into them.

One central challenge is that the path suggested leads to recommending a cautious one-small-step-at-a-time approach, which is not sufficiently bold or audacious to satisfy many advocates of interdisciplinarity. One goal of interdisciplinary work could certainly be (and, in many cases, is) to make "representational quantum leaps", that is creating entirely new, original, and innovative ways of representing phenomena by combining elements "like they have never been combined before". The more cautious goal of *de-idealisation* has its own set of challenges. Some of the more successful cases of interdisciplinarity discussed above certainly qualifies for being categorised as de-idealisation (even if their starting point may not have been deliberately idealised, but were simply an initial best approximation). I will discuss these two strategies and the continuum between them later in this chapter.

As will be evident, the framework presented in this chapter differs radically from more standard ways of construing interdisciplinarity. Viewed as integration between approaches, interdisciplinarity includes activities which would not be included on standard accounts hereof. For instance, the import of operational definition into psychiatry is not normally considered an example of interdisciplinarity. On the other hand, approach-based analysis is able to encompass everything discussed in the Interdisciplinarity Studies literature, though it is not certain, of course, that noteworthy epistemic integrations take place in all these cases.

It should be clear, though, that approach-based analysis of interdisciplinarity draws out aspects which ought to be included in any assessment of the epistemic vices and virtues of cases of disciplinary integration.

As a final important remark, in the discussion below it is assumed that integration is achieved by adding elements to a base approach belonging to one's home discipline. But of course, it would be possible to cherry-pick and combine elements more freely without relying at all on a familiar basis. This is, however, probably rarely how interdisciplinary integration proceeds. Indeed, it would require considerable effort to free oneself completely from one's educational and professional background. Indeed, funding would be considerably more difficult to obtain, if one were not affiliated with some academic milieu. In the following, then, it is assumed that there is one discipline more basic than the others involved. This means that in many cases, what is not explicitly changed will remain in the default order of the base discipline.

The Simple Duplex

As argued above, Giere's account of scientific representation provides a good foundation for understanding scientific activities in terms of representation. A very simple move, the construction of *the Giere duplex*, provides a useful backdrop for understanding the epistemic dynamics of interdisciplinarity. In this chapter, the primary discussion of epistemic aspects of interdisciplinarity will be carried out in terms of approaches. But before we focus narrowly on the integration of approaches, we must first situate our discussion in its natural environment: The integration of scientific representations.

Supposing that scientific activities can be characterised by how some specified phenomenon is represented, interdisciplinarity can be understood as the combination of two (or more) Giere-style representational relations. For the sake of simplicity, assume two distinct groups of scientists using distinct models to represent distinct aspects of the world for distinct purposes. Based on Giere's four-place-relation, the relation between pre-integration activities and their integrated state of affairs could be represented as in Fig. 7.1.

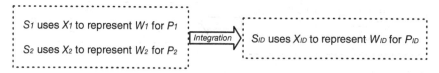

Fig. 7.1 The Giere duplex

This duplex version of Giere's representational relation leaves us with four primary dimensions along which to analyse integration, namely S, X, W, and P. Put differently, the Giere duplex provides us with four very general aspects of representation, which gives us some initial idea of different ways in which scientific activities can be integrated. As mentioned above, the beauty of the duplex is that aligns almost perfectly with the best and most nuanced characterisations of the nature of disciplines discussed in Chap. 2.

To repeat, S means 'someone' (rather than 'scientist'). S includes institutional affiliations, educational background, power relations, personal and professional ambitions, as well as other aspects of sociological interest. X is the vehicle of representation; W is the targeted (class of) phenomena; and P is the purpose of the representational enterprise.

S, X, W, and P are the parameters to study when looking for initial indications that interdisciplinarity is taking place. Publications are where we look.[1] Approaches are what we ought to look at in order to evaluate the epistemic aspects of interdisciplinarity.

In Fig. 7.2, the dotted and the dashed ellipses indicate the parent approaches, whereas the dotted-and-dashed ellipse indicate the integrated approach.

Representing the pre-integrated state of affairs of a case of interdisciplinarity in the manner of the Giere Duplex is strongly idealised in several ways.

First, it is unlikely to find a group of scientist operating with only a single vehicle of representation which they apply to only a single aspect of reality. Nevertheless, in a given (part of a) publication where a specific approach is presented, one specific vehicle is used to represent one specific class of phenomena. In case more than one vehicle is used to represent one (or more) phenomena, or one single vehicle is used to represent several

Fig. 7.2 The Giere duplex—"approached"

phenomena, this will simply mean that more than one approach is being presented.

Second, in many interdisciplinary cases integration is not restricted to the combination of elements derived from two parent approaches. Thus, in many cases it would be more adequate to represent integration as in Fig. 7.3 instead.

However, since integrating two distinct parent approaches must be the minimal requirement for interdisciplinarity, I think it is reasonable to stick to the simple duplex version for the present discussion. Once we get a handle on the challenges of the simple duplex, the move to considering more complex *"n-plex"* integration is relatively straightforward. Consequently, this idealisation is acceptable, not least since it is now made explicit that in actual cases, things will often be more complicated.

Third, the relation between X (the vehicle) and W (the target) is far less simple than displayed in Giere's four-place-relation. As discussed at length above, the relation between X and W is mediated through an intermediate layer of propositions, assumptions, and tools, and there is, therefore, not a simple, direct relation between vehicle of representation and the targeted phenomenon.

Central to the account of representation presented above is the need to make explicit *how* vehicles are used to represent phenomena. That is, what is the make-up of the intermediate layer linking vehicle and target? Which elements are retained in the integrated state of affairs, and how are they combined? One important conclusion above was that each of the elements constituting the intermediate layer are themselves candidates for relocation between approaches in interdisciplinary integration. However, in the Giere Duplex this crucial, mediating layer is black boxed (see Fig. 7.4).

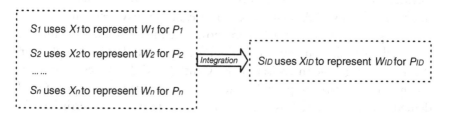

Fig. 7.3 The Giere *n-plex*

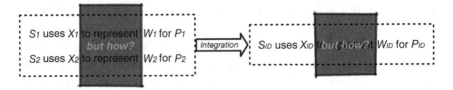

Fig. 7.4 The small black box

In spite of these imperfections and difficulties incorporated in the Giere duplex, my conclusion is that it is beneficial initially to consider interdisciplinarity in this very simplistic way. One central benefit is that the Giere duplex makes these idealisations plain for everyone to see. I will stick with Giere's basic four-place-relation for the initial discussion, before addressing the more delicate issues by means of approach-based analysis.

Starting out, then, with the (simple) Giere duplex, we have four obvious parameters for integration. Before turning to the central discussion of approach integration, I will briefly go through integration along the dimensions which are not considered to be part of the approaches.

Social Integration

Viewed in the perspective of the Giere Duplex, the first thing to catch the eye is that in most of the Interdisciplinarity Studies literature focus is almost exclusively on S-integration as discussed in some detail in Chap. 3. The next thing to notice, is that the strong focus on the social aspects of interdisciplinary integration in this literature seems to partly occlude the complexities of integrating along the X-, W-, and P-dimensions.

S-integration is first and foremost a question of unifying otherwise distinct groups of people involved in scientific representation. How might that be done? At first glance, S-integration may seem simple: The original groups must be united somehow. The involved scientists may be physically placed in the same building, perhaps complete with a plate on the door saying "The Interdisciplinary Centre of XYZ", or, less thoroughgoing, administratively placed under a common leadership (perhaps, if denied a door-sign, at least provided with a website). Nevertheless, in many cases the integrated group will be provided with some sum of money, which they are to spend on carrying out their interdisciplinary collaboration.

Expressed in a less silly fashion, there are many different ways of integrating along the S-dimension, ranging from informal collaborations involving little more than the discussion of relevant topics with people affiliated with departments other than one's own, to establishing full-fledged new institutional units including educational programs, peer structure, dedicated outlets for publication, and so on.

As indicated, S-integrations are, though far from straightforward, the aspects of interdisciplinarity most thoroughly dealt with in the literature. Put in somewhat provocative terms, there is little epistemic challenge in agreeing on which people to invite for participation. One might take this provocation a step further by claiming that when we are interested in matters epistemic, it does not really seem to matter *who* is using a given approach. Only in cases where change in the S-dimension result in change in the other dimensions as well would the S-change be interesting at all. And in this case the XWP-change would still be our primary object of study. To add one further complication, interdisciplinarity is not required to involve more than one person and does not require any changes to affiliations either. Consequently, S-integration ought not be the primary dimension of inquiry in studies of interdisciplinarity.

As discussed in Chap. 3, however, it is in many cases exclusively along the S-dimension that interdisciplinary projects are evaluated. In such cases, the way interdisciplinary projects are appraised, completely circumvent the epistemically substantial aspects of scientific collaborations. The interesting and difficult central epistemic issues in the representational relation are placed in an academic and administrative black box, as illustrated in Fig. 7.5.

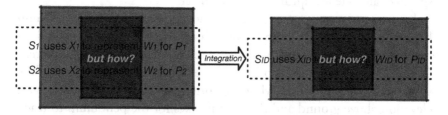

Fig. 7.5 The larger black box

This leaves the most philosophically interesting dimensions of the Giere duplex unaddressed. Consequently, according to this perspective, little if any attention is paid to epistemic aspects of interdisciplinary collaborations. In contrast to this, the overall goal of the present discussion is to remove both black boxes and uncover whatever they may conceal in all its gory detail.

Target Integration

W-integration involves arriving at a common object of study. Of the four parameters in Giere's representational relation, integration of this parameter is probably the one that seems most straightforward at first glance. Isn't it simply a question of scientists belonging to different disciplines applying their different perspectives to a common object and thereby reaching an enriched, deeper, broader, more accurate, and more nuanced appreciation of the common target? A little consideration will show that things are not that simple.

Interdisciplinarity as well as multidisciplinarity, of course, are often understood as processes in which various perspectives are applied to a common object. This raises a natural question regarding whether one can claim that an object observed in two or more ways are in a relevant way "the same" object, and further, whether what is observed must be "the same" object, for the result of the process to contribute constructively to the generation of knowledge in one or more of the involved disciplines?

Such considerations may seem unduly academic to some engulfed in the practical everyday issues of concrete science. But since it is widely accepted today that all observations are to some extent theory-laden (Hanson 1958), it is a quite obvious question whether two different perspectives are able to capture the same object.

Let us, once more, contemplate the movements of a pendulum. Kuhn has argued that if Aristotle and Galileo had observed the same pendulum in movement they would have registered something entirely different (Kuhn 1962, p. 123). Galileo would have measured period, length of the suspension and amplitude, while Aristotle would have measured weight, elevation above ground and the time it takes for the pendulum to reach rest. Would Galileo and Aristotle have had a common object?

Here we are touching upon the classical Kuhnian discussion of incommensurability. But in spite of the abundant commotion caused by Kuhn's considerations (Davidson 1974; Hacking 1983; Putnam 1975; Shapere 1966), this issue has received relatively little attention in discussions of scientific representation—and perhaps for good reason. Perhaps the focus on representation takes us to a level of detail at which the problems of incommensurability crystallises into specifications of the way in which a given model is idealised?

In fact, Kuhn's discussion of the incommensurability of the Galilean and the Aristotelian understanding of pendula seems to be a perfectly good example of an explication of two different ways of representing one and the same (type of) phenomena. In a more contemporary phrasing, it seems reasonable to argue, as Mitchell does, that differently idealised models do not target the same ideal system even if they ultimately centre on the same real world phenomena (Mitchell 2002, p. 66). Thus, there is reason to believe that incommensurability in the classic sense is less of a problem for interdisciplinarity (given adequate explication) than one might have initially supposed.

As long as one is considering concrete real world phenomena, integration of targets must simply mean to agree on a common target. It makes little sense to literally integrate the objects of interest. If one group of scientists study lions and another group study horses, it is far from straightforward how to integrate their targets. However, in cases where the target is some idealised system or for that matter some constructed class of objects (such as those suffering from a specific mental illness) literal target integration may be more easily achieved. And of course, if one group studies emotions and another group studies the brain, it can make sense to collaborate on studying the neurological manifestations of emotions.

In an approach-based perspective, pointing out the target is part of the approach. Approaches must have inbuilt indexical capacities, so to speak. Though it is relevant what targets are chosen, the most interesting aspects of how a vehicle is connected to its target is internal to the approach. Therefore, approaches are the primary focus in the approach-based approach.

Targeting a Different Target by Means of the Same Approach?

There is one further target-related issue that requires attention. What does it mean to apply the same approach to a different target? If we consider the means for pointing out the target to be part of the approach, the approach must be changed if it is to point out something different. You can hardly use the definition of 'horse' to point out lions.

At this point, we must be careful. Since according to definition all the tools and assumptions which serve to connect vehicle and target are part of the approach, it seems that one cannot use the same approach to target two different phenomena. Something must be changed, if an approach is made to target something different. Does this mean that when an approach is relocated, something is always changed at least in terms of the elements which are used to point out the target? In most cases, something will probably be changed, yes. One might in principle use the exact same approach in a different setting or context, but most cases of relocation will implicate change of (at least some details of) the intermediate layer. Of course, gaps are bridged by means of assumptions, and one may simply change assumptions about how to bridge the ultimate gap between approach and target. Still, this constitutes a (small, perhaps minimal?) change of approach.

On the other hand, there is nothing to prevent different approaches from targeting the same phenomena. The only thing required is that the parts of the approach that serve to point out the target are identical (in the relevant sense). The rest of the approaches may be entirely different.

Since the processes that point out the target are part of the approach, changes to what is pointed out should be detectable in an approach-based analysis. This, once again, indicates that it is reasonable to place approaches centre stage when analysing interdisciplinarity.

Purpose Integration

With respect to the integration of purposes, the cynic might be tempted to claim that an obvious common purpose for two groups of scientists engaging in interdisciplinary collaboration might simply be to obtain funding. The initiations of interdisciplinary collaborations are, indeed, probably sometimes motivated by the need for funding, not least since interdisciplinarity to some extent is often more or less a formal requirement for approval. But such sarcasm might nevertheless be somewhat misplaced.

In Giere's framework, funding is not a *representational* purpose of the relevant type. Rather, 'purpose' denotes the goal which the specific representational activity is intended to achieve, such as Watson's purpose of representing the physical structure of DNA (Giere 2004, p. 749).

The set of purposes to which funding belongs is a matter for economical, psychological and sociological, rather than representational, analysis. In contrast, the purposes discussed in the Interdisciplinarity Studies literature are more along the lines of "we should do a lot of interdisciplinary collaboration because then we can solve a lot of complex problems" which is yet another version of 'purposes'. Bottom line is that 'purpose' is used in many different senses, which needs to be kept apart in order to avoid confusion.

Apart from this, it is difficult to state very much in general about the integration of representational purposes other than there is little standing in the way of choosing purposes freely. The assessment of the epistemic virtues of a given approach can to a large extent be carried out independently of the purposes of the representational efforts. Indeed, the evaluation of whether some effort lives up to its purpose is dependent on a primary epistemic assessment. Again, the discussion of the involved approaches should be the primary focus.

One more thing worth considering is cases of *explorative interdisciplinarity* where the purposes of the efforts may not be clear at the outset. Giere has stated that literal truth is rarely important to scientists in practice, but "what matters is the purpose at hand" (Giere 1988, p. 78). This raises the interesting question of whether one can develop or appraise a vehicle without knowing its purpose, and whether a post hoc developed purpose is a bad thing?

Approach Integration

Rather than discussing the integration of vehicles in isolation (i.e. the X-dimension), we shall move directly to discuss integration of approaches. In the present framework, it makes little sense to discuss the X in isolation from the elements which serve to link it to its target.

One might, of course, imagine two vehicles of representation being combined, for example, in a case where one equation is inserted in another equation so as to de-idealise a parameter. However, one of the most significant conclusions I have reached so far, is that it makes little sense when analysing scientific representation to study vehicles in isolation from its target and the elements constituting the link between the two. Therefore, I will refrain from discussing vehicle-integration and focus my attention on the integration of approaches. Thus:

> Obviously, integrating scientific approaches is a tremendously complex process. There are innumerable ways in which such combinations can be made, and I do not suggest, at all, that I am able to deliver an exhaustive analysis of all possible ways of integration within the limits of this book. What I will attempt, though, is to provide some illustrative exemplifications and point out some of the causes of the great complexity of such integrations.

In many cases, it is possible to distinguish the vehicle of representation clearly from the elements mediating its connection to the target, simply since the vehicle is often proudly presented in publications as *the* model or *the* theory. Functionally, however, there are but arbitrary ways of distinguishing the vehicle itself from the elements mediating its connection to the target phenomena.

Consequently, in the following discussion of approach integration, the intermediate layer and the vehicle of representation will be considered as a unity, while the target is considered in isolation. One reason why I think this makes sense in the analysis of interdisciplinarity, is that it is possible to relocate elements between the mediating layer and a vehicle of representation or using what is used as a vehicle in one context as an element of the intermediate layer in another context. On the other hand, it does not seem to make much sense to say that one transfers some theoretical element to some real-world phenomena. Of course, a theoretical element may be targeted in an act of representation, but that is a different story.

As discussed above, I construe the connection between vehicle and target as constituted by an intermediate layer consisting of more or less explicit, more or less taken for granted, assumptions and (conceptual) tools of various kinds.

The function of each element is to translate inputs into forms (e.g. concepts or figures) which can be processed further by means of other tools, until the connection to the parameters of the vehicle of representation can be established. Some of the tools involved may be literal tools such as scientific instruments, others are mathematical tools such as statistical methods. Yet others are conceptual tools such as reasoning strategies or definitions. A subgroup of tools involved are those I call 'propositional algorithms' and define as (more or less explicitly stated) sets of rules for carrying out conceptual operations.

Involved in the intermediate layer are also sub-representations, including what is often referred to as 'data models'. Data models are ordered groups of data represented in ways appropriate for a certain purpose. Data models can be analysed in order to derive inputs to feed other elements further down the representational chain. The results of processes of measurement and categorisation also count as sub-representations (as discussed in Chap. 1).

As a highly stylised illustration of how to construe the intermediate layer and its role in linking vehicle and target, please consider Fig. 7.6. The pointy shape demarcates the approach, whereas the black line illustrates the series of steps of the intermediate layer which serve to link the vehicle to the chosen target. Each of the zigzags of the black line marks the effect of an element of the intermediate layer.

Obviously, this is not a true-to-life picture of scientific representation, which hardly proceeds along a series of steps in as simple a way as portrayed here. It would be more adequate to represent the intermediate layer as a complex network of interacting elements, but this would also be a far less accessible illustration. Indeed, representational activities often do involve observing in a certain way, and the results of observation are often processed through definitions of a certain kind, and so on. For present purposes, this corresponds sufficiently well to the activities of, for instance, the imaginary social psychologist studying the "pet-effect" as discussed in Chap. 1.

As discussed at length already, integrating two or more approaches is certainly a very complex undertaking, even in a very simplified version. In Fig. 7.7, the elements of the parent approaches (vertical and horizontal stripes respectively) are combined in the purple approach.

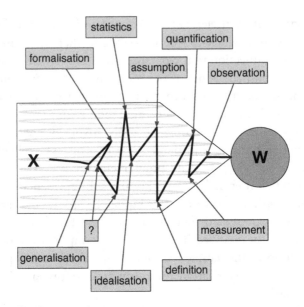

Fig. 7.6   A stylised approach

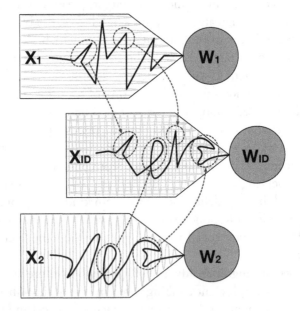

Fig. 7.7   Integrating approaches

There are a number of different complications which must be pointed out.

1. Importantly, all the elements of the parent approaches are candidates for being reused in an integrated approach.
2. Equally important, nothing stands in the way of using the elements differently in the integrated approach compared to their origin.
3. No elements of the parent approaches are guaranteed a place in the integrated approach. Even the most fundamental assumptions of parent approaches can be (more or less deliberately) disregarded in the integrated approach.
4. In many cases, combining elements from various sources requires additional elements (often in the form of assumptions) not found in the primary parent approaches. These may then be developed specifically for the integrated approach or may be imported from somewhere else. These must be identified and their function must be determined.

A further, very important complication: Certainly, the origin of the specific combined elements is far from always explicated. To complicate matters even more, in most cases the elements will not be imported directly from the original source, or even from a source in near disciplinary proximity to the original application of the theoretical element. A prime example hereof could be the use of statistical tools as discussed in the previous chapter. When psychologists make use of the ANOVA, they are rarely interested in its original application. Indeed, in psychology textbooks statistical tools are most often presented anonymously, that is, without reference to the originator or the original context of a specific tool. It has been argued that this can have the unfortunate side-effect of strengthening the impression of neutrality and of statistics as constituted by a bundle of more or less "given truths" (Gigerenzer 2004, p. 589).

The same pattern will be seen in relation to other tools such as semi-structured interviews, which will play a role in the case study discussed in the following chapter. In the eyes of the non-expert, the semi-structured interview may appear as a firmly established specific method, when in fact there is room for almost endless variation regarding how to carry out such interviews (Kvale and Brinkmann 2009).

To a large extent, Giere draws his examples for analysing representational activities from science textbooks. This provides for an interesting contrast to our concern with interdisciplinary scientific work. Since interdisciplinary projects are almost by definition ground-breaking, their conclusions cannot as of yet have sedimented into textbooks. Often people engaged in interdisciplinary projects will be hesitant in drawing on textbooks as well, since textbooks are (reasonably) considered less advanced. Further, there is the interesting complication that drawing on textbooks will provide you with what is often insufficient insight into an idealised version of "the real thing".

An approach to representation as inclusive as advocated in this book results in an enormous diversity with regard to types of vehicles, structure of the representational relation, types of explanations used, degrees of universality or individuality, level of specificity, and so on. This diversity inspires a lot of questions. One obvious question is whether there are any restrictions as to which types of approaches are possible to integrate? It is difficult to see principled limitations here. There is nothing that hinders the construction of approaches that combine elements from chemistry, biology and, perhaps, psychology or literature theory to ascribe agency to molecules or bacteria, for instance. Obviously, there are relevant questions of adequacy, validity, power of prediction and so on. But these are questions related to the evaluation of the resulting approaches, not to the possibility of constructing them (Collin 2011; Latour 1988).[2]

So, at the outset there are no principled limitations as to which parent approaches it is possible to combine. Whether the combined approaches are distant or proximate or somewhere in between, there is little inherent resistance (if any) in the process of combining theoretical elements into some representational relation. Representation is certainly constructive in this sense: Representation is something you do, and at the outset you are free to do (i.e. represent) as you please. It is not until we stringently and systematically start checking the mediated fit between vehicle and target that we start to realise to which extent a particular way of representing is epistemically successful.

The integration of two or more approaches can thus appear to be quite easy at first glance. But creating a representational approach with a good fit is a tremendously challenging process. There are numerous ways to combine elements—and most of these will show poor results (from an epistemic point of view).

The Method

If we are to evaluate a case of interdisciplinarity along the lines suggested, the first step, once we have decided on a specific publication to analyse, is to single out the parent approaches and gauge the distance between them. The next step is to determine which elements from the parent approaches are reused in the integrated approach as presented in the publication and how they are combined. The third step is to determine which functions the elements serve in the integrated approach and whether there are newly developed elements involved, or elements imported from elsewhere than the primary parent approaches. In all cases, with respect to all imported elements the extent to which they are transformed, idealised, or assigned new roles should be determined.

Spelling this out in as much detail as we are able will provide us with some insight into the structure of the integration taking place. Recalling the suggestion to consider disciplines as bundles (of bundles) of approaches, one must keep in mind that some aspects might be more or less taken for granted characteristics of an entire bundle (or even a bundle of bundles). Issues taken for granted in an entire bundle may not be explicitly stated in the presentation of a particular approach.

The final and most demanding step is to evaluate how the relocated elements fare in their new contexts and, on this background, how the interdisciplinary approach fares. The more proximate the parent approaches are, the easier one can expect the task to be, because alterations are likely to be less radical. As distance increases, the demands of integration, as well as the analysis hereof, increases as well. This is, however, only a rule of thumb. One should always beware of important changes even in proximate integration.

Transferring Vehicles

The obvious place to start a dissection of an interdisciplinary case is by focusing on vehicles of representation. That is, one must determine what is used as a vehicle of representation in the integrated approach, and how this vehicle relates to the vehicles of the parent approaches.

The most high-profiled examples of transferral of theoretical elements between approaches are the relocation of entire vehicles of representation which are assigned to new targets. Especially in the case of mathematical models this may appear to be straightforward. Due to the abstract nature of mathematical models, (apparently) little if anything tie such models to their original applications. Thus, nothing (apparently) prevents one from transferring, for instance, equations developed in physics and using them to represent some phenomena in economics. For example, Philip Mirowski has argued that modern economics came into existence by importing mathematical structures from energetics into social science (Mirowski 1989).

Indeed, economists appear to be large-scale importers of mathematical models. Michael Weisberg has discussed one example, in which Richard M. Goodwin has imported one of Weisberg's favourite analytical specimens, the Lotka-Volterra model of predator-prey-interactions, into economics (Goodwin 1967; Weisberg 2013, p. 77).

Goodwin originally presented this idea as "a starkly schematized and hence quite unrealistic model of cycles in growth rates" (Goodwin 1967, p. 54). In his reinterpretation, Goodwin replaces predators and prey with wages and profit respectively. This raises the interesting question of to which extent a given mathematical structure used as vehicle of representation in a new context retains any connection to its original context. Does anything but history connect a mathematical equation to the contexts in which it was previously used? There is no clear answer to this question.

On the one hand, if viewed in isolation the mathematical structure is not tied to its original target by anything but the way it was used. And indeed, there are probably many examples in which some equation fit better in a new context.[3] On the other hand, in many cases it is not just the equation which is imported. In many cases, when importing a certain equation (or other theoretical element) one (more or less deliberately) imports other aspects as well. For instance, for the shareholder there is an element of moral support in considering one's hard-earned profit to be pursued by employees driven by their insatiable demands for higher wages.

A further interesting consideration is whether there are special circumstances in economics which makes it especially tempting to import mathematical structures. Indeed, there might be. Throughout economic

history theorists have been attracted to the combination of strong idealisation and quantification. Think for instance of the idealisations required
in order to be able to provide numerical measures on a common scale for
divergent types of value. Such procedures are required for, among many
other things, the mathematical analysis of opportunity costs. Other frequently discussed examples of central and important economic idealisations are the perfectly transparent free market as well as its inhabitants.
Homo economicus is the strongly idealised model of man sometimes considered to capture economic behaviour in condensed form.[4]

Once something is quantified, it is tempting to feed the numerical
data into equations—and why not use off the shelf equations? But as
discussed in Chap. 5, while strong idealisation makes it easier to construct mathematical systems which fit nicely together, one might pay a
high price in the currency of fidelity.

The pattern of importing and re-construing a vehicle of representation
in its entirety is hardly the way most interdisciplinary activities proceed.
It is rarely the case that one party provides the vehicle of representation
while the other party provides the intermediate layer (and the target?). We
therefore need a more fine-grained analytical approach to capture more of
what is going on. Luckily, in our present framework, we are by no means
restricted to analysing the transfer of vehicles of representation.

Inserting Elements of Approaches as Parts of Vehicles

The next option to consider is whether an existing vehicle of representation has been changed as a consequence of the integration. This may, for
instance, arise through the transferral and insertion of elements of one
approach as part of a vehicle of representation.

In some cases, (elements of) one approach may be able to cast light on
an underdeveloped aspect of another act of representation. A good example can be found in the history of evolutionary biology, where, as is well
known, Darwin had postulated inheritance and variation as central parts
of his theory of natural selection without being able to explain the underlying mechanisms. Indeed, Mendelian geneticists argued that Darwinians

had no adequate basis from which to explain changes resulting in specia-
tion (Burian 2005, p. 105).

When Dobzhansky published *Genetics and the Origin of Species* (1937),
he very successfully provided the missing link.[5]

> It was Dobzhansky's virtue to bridge the communities of naturalists and
> geneticists, to find a methodology for doing experimental work with natu-
> ral populations, and to write a book that so formulated the issues that, after
> much dispute, both communities were, by and large, persuaded. [...] In
> *Genetics and the Origin of Species* [Dobzhansky] synthesized his findings
> and those of many others to provide an integrated account, consistent with
> genetics, of the origin of species as an extrapolation of the microevolution-
> ary changes that the naturalists had described in detail. (Burian 2005,
> p. 106)

No doubt the introduction of genetics added considerable strength,
nuance, and accuracy to evolutionary biology. But even though the synthesis of genetics and evolutionary theory has been tremendously successful, lots of unanswered questions remain. Importantly, it is worth noticing
that genetics as well as the naturalist approach are both perspectival in the
sense that genetics emphasises microscopic structures and gene frequen-
cies while the focus of naturalism is on the role of possessing different
phenotypes. Even though, I believe, most will agree that the evolutionary
synthesis is superior to each of its parent approaches, 80 years of careful
efforts has not yet fully solved the inherent controversies. Topics such as
what level of abstraction (genes, phenotype, species, ecosystem, and so
on) might be the most proper level of selection remain controversial even
today. Apparently, not even the most impressive of integrative successes
deliver instant gratification.

Relocating Elements of the Intermediate Layer

The next task is to determine which elements of the intermediate layer of
the parent approaches are reused in the integrated approach. In this case,
one must distinguish between two classes of relocation. The first is the
transfer of elements to serve the same function they did in the parent

approach. The other is, as discussed extensively above, situations in which elements are used differently in the integrated approach than in the parent approaches from which they are imported.

When a tool is relocated, there are a number of aspects which one might inquire into in order to determine to what extent it is used similarly to the original context. Most importantly, where it is located in the line of reasoning, the nature of the inputs it is fed, and what its outputs are used for further down the line.

As discussed earlier (the treatment of *Insight A* and *B* in Chap. 3) there are no a priori reasons to assume that some element will be more or less useful simple for being used differently. However, in cases where a certain tool has a long track record it has often been rigorously scrutinised by people with specific, relevant expertise. Therefore, one might suspect that tools have often been optimised and calibrated to serve their functions reasonably well in their original contexts. If this is the case, one should expect to find a comprehensive literature on the development and use of the particular tool. In such literature, details about the utility and special requirements of the tool in question will often be spelled out. Indeed, this was the case in the examples I discussed in the previous chapters. Non-experts in new contexts, on the other hand, are less likely to check in deep detail the viability of, for instance, some statistical tool or a certain unusual type of definition.

Examples of tools moved to a different function in a new context include operational definitions and the ANOVA discussed in detail in the previous chapter. Figure 7.8 illustrates moving some element from carrying out a very basic function to a different location. This is similar to the discussion of operational definition in psychopathology. But the patterns exhibited in the examples discussed above do not exhaust the possibilities, of course. One might transfer and mix up techniques, instruments, tools (literal as well as mathematical and conceptual), ways of observing, ways of doing experiments, ways of collaborating, ways of communicating. The possibilities are innumerable.

It would be a foolish exercise to attempt to go through each and every possibility here. The important thing is to be cognisant of the general pattern of transferring elements that serve various functions in connecting vehicles of representation to their targets. One must pay close atten-

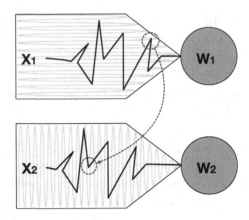

Fig. 7.8 Relocation of elements

tion to alterations in the function elements are used to serve and the inputs they are fed in the new interdisciplinary context. One must further pay close attention to relocational idealisations as discussed in Chap. 6.

The discussion of ceteris paribus clauses in Chap. 6 emphasised that we should not expect tools to work universally. When assessing the viability of relocating a specific tool one should pay close attention to its specific requirements and restrictions. Some tools will only operate well within narrow limits, in other cases restrictions may be more loose. When measuring gravity waves (Abbott et al. 2016) or slowing the speed of light (Hau 2011) there is very little room for play or variation, especially in the experimental set-up. In other settings, such as when normal functioning capacities for categorisation are used to decide whether an elderly person owns a dog or a canary, requirements are considerably laxer. One will consequently need to assess the strictness of ceteris paribus clauses on a case by case basis.

Two Strategies

I believe there are two general strategies which may direct interdisciplinary activities. One can be characterised as the attempt at an asymptotic approach towards truth (likeness) by carefully clearing up imprecisions resulting from idealisations and approximations. The other is less careful and characterised by a more radical innovative attitude (perhaps by making bold conjectures). The latter of these is far more difficult to analyse and, one could fear, likely to raise more questions than one would be able to answer.

Strategy #1: De-idealisation

Most of the successful instances of interdisciplinarity I have been able to identify proceed in the manner of de-idealising. The ideal of de-idealisation, so to speak, can be illustrated by returning to an earlier used example.[6] As discussed at length above, the equation of the simple pendulum (see Fig. 7.9) is highly idealised. One of the ways in which the equation is simple is that it only captures pendula with a very small angular amplitude (Θ) (i.e. the angle of the rod measured from the downward vertical). Indeed, the angular amplitude is required to be so small that it is not even represented in the equation.

The consequence is that the equation for the period of a pendulum is only reasonably accurate with regard to angular amplitude of a few degrees. The treatment of the large amplitude pendulum, however, is much more complex.

One might de-idealise the equation of the simple pendulum by inserting an additional mathematical structure. One can approximate a solution to this problem by a series for which the first terms would be as in Fig. 7.10.

$$p = 2\pi \sqrt{l/g}$$

Fig. 7.9 The simple pendulum once more

$$p = 2\pi \sqrt{l/g} \left[1 + \frac{1}{16}\theta^2 + \frac{11}{3072}\theta^4 \dots \right]$$

Fig. 7.10 The less simple pendulum

It should be obvious that as Θ approaches 0, the influence of this addition becomes more and more negligible. Nevertheless, adding an approximation to angular acceleration to the equation of the simple pendulum is a way of *de-idealising* it. By this procedure we get a less idealised equation which is a better approximation of the real thing (or actually in this case enables the de-idealised vehicle to capture a larger part of the potential behaviours of real pendula).

According to Weisberg, we idealise for pragmatic reasons, that is, because we lack the computational power or insights required to compute the problem at hand in all its gory complexity. Analysing an idealised version of the issue at stake, however, may provide us with some of the insight required to take the analysis to the next step by developing a carefully *de-idealised* version. Once we get a handle on some isolated aspect of a complex problem, this might reveal ways forward towards handling the issue with more context included.

McMullin discussed de-idealisation in his classic paper on Galilean idealisation:

> [...] [M]odels can be made more specific by eliminating simplifying assumptions and 'de-idealizing', as it were. The model then serves as the basis for a continuing research program. [...] If simplifications have been made in the course of formulating the original model, once the operations of this model have been explored and tested against experimental data, the model can be improved by gradually adding back the complexities. Of course, this requires a knowledge of how that particular 'complexity' operates. (McMullin 1985, p. 261)

Mitchell has discussed similar kinds of representational improvement in terms of "[...] a strategy of asymptotic approach to a non-idealized representation of the [target] system" (Mitchell 2002, p. 65).

In many interdisciplinary activities one can detect a somewhat naïve aspiration of de-idealisation. The naïve thought can be expressed in terms like "By adding an additional theoretical element to our construal of our target we combat over-simplifications and gain in actual similarity between vehicle and target". However, it is somewhat paradoxical that in many cases the imported theory is indeed itself idealised as part of its relocation. In some ways, then, idealisations are used as a means for interdisciplinary de-idealisation.

Good idealisations are carefully worked out step by step for specific purposes in specific contexts. The same holds for de-idealisation as described in the McMullin quotation just above. As stated several times, there are no good reasons to assume that the combination of two or more approaches automatically adds up to anything like an asymptotic approach to a non-distorted, de-idealised representation (Mitchell 2002; Wimsatt 1987).

Strategy #2: Bold Conjectures

As should be evident by now, according to the present analysis interdisciplinarity is not expected to be an easy shortcut to epistemically sound scientific originality or innovation. On some accounts, however, the ideal of interdisciplinarity seem to be a variation on the bold leaps once endorsed by Karl Popper (1963). However, unless issues such as those pointed out in the present discussion are taken into account, interdisciplinarity is in danger of ending up being all about bold conjectures while completely ignoring the significant part about refutation.

The bold way of combining distant approaches could also be construed as explorative interdisciplinarity. Let's integrate something and see what happens. If, however, one is inclined to consider interdisciplinarity as inherently good, one might be in danger of underestimating the considerable epistemic risks related to such representational leaps into the unknown.

Other than these scarce remarks, it is difficult to say anything in general about the bold strategy, except that it is likely to require considerably more effort to assess compared to the de-idealising strategy. Most interdisciplinary projects will probably be located somewhere on a continuum between the extremes constituted by strategy #1 and strategy #2. Nevertheless, it may be worth considering which of the extremes some specific interdisciplinary case most closely resembles.

Summing Up

The suggestion of this chapter is to turn standard methods for analysing interdisciplinarity inside out and focus exclusively on what is usually black boxed. This means bracketing (or black boxing) everything but approaches and targets. This approach does not constitute an exhaustive analysis of interdisciplinarity, of course, but it captures the central epistemic aspects of such activities.

The Giere duplex is only useful for a coarse, initial analysis of actual cases of interdisciplinarity. For the purpose of epistemic evaluation, we must maintain the focus on approaches as developed above. Still, it is worthwhile to start out a representation-based analysis of a case of interdisciplinary science by drawing out the dimensions emphasised by the Giere duplex. This will constitute useful contextualisation when moving on to the more detailed approach-based analysis.

The suggestion is, then, to first and foremost focus on approaches when addressing issues related to the epistemic benefits of interdisciplinarity. I do not suggest (at all) that one should ignore the social aspects of interdisciplinarity. When addressing troubling issues such as degenerating core sets or the lack of peers sufficiently competent to do qualified reviews, social matters should not be ignored. But when it comes to analysing the extent to which a certain interdisciplinary activity has led to epistemically beneficial results, approaches are the heart of the matter.

One central assumption in approach-based analysis of interdisciplinarity is that one should not trust the prima facie authority provided by imported tools, propositional algorithms, assumptions, vehicles, or what have we. Any act of representation must be evaluated afresh. This is very laborious and requires a stringent and systematic attitude—but it is the way forward.

Once having determined as closely as possible what is being used as vehicle of representation, the next step is to identify as many relevant elements of the intermediate layer as possible. One cannot evaluate the vehicle in isolation from the intermediate layer, not least since a significant part of the integration might take place there.

When evaluating the individual elements of the intermediate layer, the following focal points are useful:

- What is the representational context in which the element is put to use?

  – How does this context differ from the context in which the element was used in the parent approach?

- What inputs are fed to the element?

  – How do these inputs differ compared to the inputs in the parent approach?

- What are the requirements of the element?

  – How does the present use live up to these requirements, and how does this differ from the parent approach?

- What might the element be expected to deliver?
- How well does the element perform in the parent approach?

  – as well as, perhaps, in its original or optimal context?

- How well can the element be expected to work in the function it is assigned in the integrated approach?

One might think that it is unnecessary to consider how well a certain element performed in the parent approach since "Insight A" and "Insight B" above gave us reason to think that we need to evaluate its performance from scratch anyway. However, as mentioned above, since many tools are developed by experts in a specific area these will often have engaged in thorough discussions with their peers about the specific tools. Even though these discussions do not provide the final word on how a tool will fare in an interdisciplinary context, expert discussions may certainly point out issues relevant to consider. You don't have to be an expert on statistics to appreciate some of the challenges pointed out in the statistical literature on ANOVA, for instance.

In many cases, unfortunately, it is impossibly difficult to figure out what is actually going on. What elements are mixed up? Where do the elements come from? Anonymisation plays a crucial role here. Because if insufficiently explicated contributions are anonymised and, for instance, only referred to as 'psychoanalytical', matters may be impossibly difficult to figure out.[7]

Focus in the literature on interdisciplinarity has neglected epistemic aspects. The way to make up for this shortcoming is to pay especially close attention to relocation, including idealisations and other alterations of vehicles and mediating theoretical elements. A focus on vehicles and elements of the intermediate layer and the transformations they go through will reveal a lot about the viability of specific interdisciplinary approaches. In this way, approach-based analysis will constitute a fruitful alternative to standard approaches to the study of interdisciplinarity.

I believe, thus, that a number of the shortcomings of Giere's representational relation can be handled by making some of the adjustments and additions I have already discussed above. The problems related to underdetermination of the relation between X and W can be properly handled by including discussions of assumptions, tools, and propositional algorithms. The problems related to analysing interdisciplinary representation in terms of groups of scientist using only a single model to represent a single target may be handled reasonably by means of my notion of "approach" as discussed above.

The result is, I believe, a method capable of drawing out numerous interesting aspects of interdisciplinary science highly relevant to the epistemic assessment of interdisciplinary activities.

Notes

1. Of course, a large part of the interdisciplinary activities may take place prior to the first publication of an interdisciplinary research project. Approach-based analysis is primarily a method for retrospective assessment of interdisciplinary science. However, the pointing out of potential difficulties to avoid means that the method has considerable utility in the early design phases of interdisciplinary science as well.

2. I thank one referee of an early version of (Hvidtfeldt 2017) for bringing to my attention (by criticising me for using imagined examples of metaphorical use of psychology to understand bacteria) that it is too subtle to simply support that claim with a reference to *The Pasteurization of France* by Bruno Latour. With an enchanting touch of modesty Latour refers to this book as his "Tractatus Scientifico-Politicus" (Latour 1988, p. 7). If one consults this seminal and explicitly interdisciplinary work within

actor-network theory, one will see that the ascription of agency to even the tiniest of non-human "actors" is (1) apparently not meant metaphorically, and (2) not something that I have imagined!

3. This certainly does not imply that mathematical structures are neutral, understood as non-perspectival (as discussed in the previous chapter).

4. Mill introduced the concept of "*homo economicus*" in his treatise on method in a deliberate attempt at making economics more scientific. Ironically, he was confident that no political economist would ever be "so absurd as to suppose that mankind are really thus constituted" (Mill 1836, p. 322). Interestingly, there is a tension between two possible interpretations of the idealisations involved in the construction of the model of economic man. According to one interpretation, the idealisation emphasises the "essential" characteristics of human economic behaviour. Since they are essential, they are common to all economic agents, and the model thus applies everywhere—in other words: it is generalisable. If we encounter differences between instances of actual behaviour, they are accidental and ought (ideally) to be negligible. The other interpretation reaches the opposite conclusion: The model of economic man applies nowhere, since persons thus characterised in fact do not exist (Morgan 2006, p. 6). In economics, it often seems that the first interpretation is accepted, even though the latter is probably much closer to actual fact.

5. Pardon the (deliberate) pun.

6. You guessed it: The simple pendulum!

7. Since there are no generally agreed upon psychoanalytic approaches to rely on.

References

Abbott, B.P., et al. 2016. Observation of Gravitational Waves from a Binary Black Hole Merger. *Physical Review Letters* 116 (6): 061102. https://doi.org/10.1103/PhysRevLett.116.061102.

Burian, Richard M. 2005. *The Epistemology of Development, Evolution, and Genetics.* New York: Cambridge University Press.

Collin, Finn. 2011. *Science Studies as Naturalized Philosophy. Synthese Library Studies in Epistemology, Logic, Methodology, and Philosophy of Science.* Dordrecht; New York: Springer.

Davidson, Donald. 1974. On the Very Idea of a Conceptual Scheme. *Proceedings and Addresses of the American Philosophical Association* 47: 5–20.

Dobzhansky, Theodosius. 1937. *Genetics and the Origin of Species, Columbia Biological Series.* New York: Columbia University Press.

Giere, Ronald N. 1988. *Explaining Science: A Cognitive Approach, Science and Its Conceptual Foundations.* Chicago: University of Chicago Press.

———. 2004. How Models are Used to Represent Reality. *Philosophy of Science* 71 (5): 742–752. https://doi.org/10.1086/425063.

Gigerenzer, Gerd. 2004. Mindless Statistics. *The Journal of Socio-Economics* 33: 587–606.

Goodwin, Richard M. 1967. A Growth Cycle. In *Socialism, Capitalism and Economic Growth*, ed. C.H. Feinstein, 54–58. Cambridge: Cambridge University Press.

Hacking, Ian. 1983. *Representing and Intervening: Introductory Topics in the Philosophy of Natural Science.* Cambridge, UK; New York: Cambridge University Press.

Hanson, Norwood Russell. 1958. *Patterns of Discovery; An Inquiry into the Conceptual Foundations of Science.* Cambridge, UK: University Press.

Hau, Lene Vestergaard. 2011. Quantum Optics: Slowing Single Photons. *Nature Photonics* 5: 197–198.

Hvidtfeldt, Rolf. 2017. Interdisciplinarity as Hybrid Modelling. *Journal for General Philosophy of Science* 48 (1): 35–57.

Kuhn, Thomas S. 1962. *The Structure of Scientific Revolutions.* International Encyclopedia of Unified Science: Foundations of the Unity of Science V. 2, No. 2. Chicago: University of Chicago Press.

Kvale, Steinar, and Svend Brinkmann. 2009. *Interviews—Learning the Craft of Qualitative Research Interviewing.* Thousand Oaks, CA: SAGE Publications, Inc.

Latour, Bruno. 1988. *The Pasteurization of France.* Cambridge, MA: Harvard University Press.

McMullin, Ernan. 1985. Galilean Idealization. *Studies in History and Philosophy of Science* 16: 247–273.

Mill, John Stuart. 1836. On the Definition of Political Economy. In *Collected Works of John Stuart Mill: Essays on Economics and Society*, ed. J.M. Robson. Toronto: University of Toronto Press.

Mirowski, Philip. 1989. *More Heat Than Light: Economics as Social Physics, Physics as Nature's Economics, Historical Perspectives on Modern Economics.* Cambridge; New York: Cambridge University Press.

Mitchell, Sandra D. 2002. Integrative Pluralism. *Biology and Philosophy* 17 (1): 55–70.

Morgan, Mary S. 2006. Economic Man as Model Man: Ideal Types, Idealization and Caricatures. *Journal of the History of Economic Thought* 28 (1): 1–27.

Popper, Karl. 1963. *Conjectures and Refutations: The Growth of Scientific Knowledge*. London: Routledge.

Putnam, Hilary. 1975. The Meaning of "Meaning". In *Philosophical Papers*, vol. 2. Cambridge: Cambridge University Press.

Shapere, Dudley. 1966. Meaning and Scientific Change. In *Mind and Cosmos; Essays in Contemporary Science and Philosophy*, ed. Robert Garland Colodny, 41–85. Pittsburgh: University of Pittsburgh Press.

Weisberg, Michael. 2013. *Simulation and Similarity: Using Models to Understand the World. Oxford Studies in Philosophy of Science*. New York: Oxford University Press.

Wimsatt, William. 1987. False Models as Means to Truer Theories. In *Neutral Models in Biology*, ed. Matthew H. Nitecki and Antoni Hoffman. New York: Oxford University Press.

8

Phenomenology Imported with EASE

In the preceding chapters, I have developed and argued in favour of a representation-based method for capturing and assessing a number of epistemic aspects of interdisciplinary science. This chapter will provide, if not a full analysis, at least a quite detailed case study utilising the framework developed above. The case study will focus on an explicitly interdisciplinary research project within psychopathology: the so-called EASE-project (Evaluation of Anomalous Self-Experience) developed under the leadership of Professor Josef Parnas at the University of Copenhagen. The main focus will be on one approach of this project, which is presented in an article by Julie Nordgaard and Josef Parnas entitled "Self-disorders and the Schizophrenia Spectrum: A Study of 100 First Hospital Admissions" published in *Schizophrenia Bulletin* (2014). In the following, I will refer to the approach of this article as 'NP2014'.[1] I will further draw on a number of other publications within the EASE framework in order to close in on some aspects of NP2014 which are not explicitly addressed in the article in question.

NP2014 is one approach out of the bundle of approaches which collectively constitute the EASE project. One could certainly argue in favour of analysing another approach from this bundle. But the focus on

© The Author(s) 2018
R. Hvidtfeldt, *The Structure of Interdisciplinary Science*, New Directions
in the Philosophy of Science, https://doi.org/10.1007/978-3-319-90872-4_8

NP2014 is chosen since I believe that the particular specimen exhibits a number of interesting aspects and, due to its recent publication, represents a fairly mature instantiation of EASE.

While developing the research which eventually led to this book, I was employed at the psychiatric research facility where most of the efforts in EASE have been carried out. This includes the training of psychiatrists in using the methods developed in EASE. As a consequence of my employment at that facility, I have participated in two EASE-courses (a basic three-day course and a two day "advanced workshop"). This means that I am by now a certified EASE-practitioner.[2] Due to this employment and previous work related to the philosophy of psychiatry and psychopathology, I have gained considerable insight into the specifics of EASE as well as the structure and dynamics of psychiatry and psychopathology in general.

Having stated these opportunistic factors outright, let me also state that I do find psychiatry and psychopathology to be especially interesting domains to study with regard to interdisciplinarity. I will discuss why in the next section. Further, the EASE project is interesting in this context since it makes use of propositional modelling in combination with various tools for translating qualitative representations into quantitative measures. This gives the reader good opportunity to assess whether these aspects are captured convincingly by approach-based analysis.

EASE is firmly situated within psychiatry. Indeed, all the publications related to EASE are published in psychiatric journals or edited volumes focused on psychiatric or psychopathological matters. I will therefore treat EASE as basically a psychiatric/psychopathological project. Non-psychiatric elements of the integrated approach will be considered as imported from "alien" approaches into the psychiatric setting.

For this reason, this chapter starts out by outlining the context of psychiatry and psychopathology in general and the general target of EASE, namely schizophrenia. Against this background, the EASE-project and its general motivations are presented. This leads to the main analysis of NP2014. As should be plain by now, the analysis will involve identifying parent approaches, assessing the distance between them, identifying vehicles of representation, identifying targets, determining which tools, assumptions, and algorithms are combined as well as their function in

parent and integrated approaches, and identifying idealisations, reloca-
tions, distortions, and perspectives—to the extent possible. This will
eventually lead to interesting conclusions about the epistemic merits of
the interdisciplinary EASE-project.

Importantly, the purpose of the presented analysis is not to pass final
judgment on EASE or NP2014. Rather, the point is to demonstrate how
approach-based analysis is capable of drawing out interesting aspects of
interdisciplinary activities.

I do believe that the case study presented in this chapter displays some
considerable virtues of approach-based analysis of interdisciplinary activities. Even though the exact details for all aspects of the integration in
NP2014 cannot be discerned, the analytical tools developed above prove
their worth, I believe. Indeed, in this context, making lack of clarity
explicit is a virtue in itself.

The analysis below will bring forth interesting aspects of how to evaluate the EASE-project qua interdisciplinary research. It will highlight
some noteworthy qualities of the EASE-project as well as point out some
difficulties which are yet to be overcome. Some of the difficulties revealed
can best be described as built-in constraints of current psychiatry. For
instance, it makes little sense to state that self-disturbances are the fundamental characteristic of a category of interest, and then pick out the
members of this category without taking self-disturbance-related issues
into account at all. However, it is a basic requirement of psychiatric
research that the latter method of categorisation is used. Such constraints
certainly have the potential to stop many innovative approaches in their
tracks. Further, the analysis will show how some central elements involved
in the integration are distorted almost beyond recognition. Other elements, which are considered more or less indispensable in the parent
approaches, are left out of the integrated approach. This causes considerable concern when viewed in the present perspective.

Since EASE is based on disciplines, each of which incorporates vast
literatures, an exhaustive analysis is impossible to achieve here. Nevertheless,
it is possible within the present framework to reach interesting conclusions. Indeed, a tool for assessing interdisciplinary activities would be of
little utility if it required book-length treatments of each case evaluated.
Luckily, it appears that approach-based analysis does no such thing.

As noted above, the first step will be to contextualise the discussion within the general framework of psychiatry and psychopathology, before moving on to a presentation of the EASE-projects.

So, What is Psychiatry and Psychopathology?

Psychiatry is the medical specialty that deals with the diagnosis and treatment of mental illnesses, emotional disturbances, and (detrimental) abnormal behaviour. 'Disorder' is a central term within contemporary psychiatry and means something like "a disruption of normal mental functions". It is not sharply de-lineated what qualifies as a disorder. This should not come as a surprise, however, since neither "normal" nor "mental" are clearly defined concepts (Murphy 2006, p. 53 ff.).

Psychopathology, on the other hand, denotes the study of the nature and aetiology of the phenomena psychiatry attempts to handle. Psychiatry and psychopathology ought to be closely interconnected, one would think. Unfortunately, this is not the case to the extent one might prefer. For instance, as discussed above (in Chap. 6) in relation to operational definition, contemporary psychiatric diagnostics is deliberately (to the extent possible) cut off from theories about causation as well as from the subjective experience of patients (American Psychiatric Association 1980, 1994, 2000, 2013; World Health Organization 1992).

The category of mental disorders covers a vast range of diverse ways of being psychologically (in a broad sense) and emotionally troubled. At the one end of the spectrum, there are people suffering from mild emotional disturbances, for instance, certain mild forms of depression, who, with or without treatment, are likely to quickly make full recovery. At the other end are, for instance, cases of schizophrenia, which are often chronic and accompanied by significant cognitive, social and emotional dysfunctions.

In spite of internal disagreements, it is not unfair to say that psychiatric and psychopathological research is largely dominated by the so-called *medical model*. According to the medical model, mental disorders are manifestations of dysfunctions of some sort. One may distinguish weak from strong interpretations of the medical model. A weak interpretation is not committed to dysfunctions of a specific type, and explanations can draw on psychological, social, as well as biological elements. A strong interpretation, on the other hand, requires explanations in terms of anatomical pathology. More specifically, it is committed to causal explanations in terms of neurological abnormalities. In plain terms, if you are mentally disordered, something must be wrong with your brain. Unfortunately, very few mental disorders have known organic causes. And indeed, in cases where organic causes are determined, there is a tendency for disorders to migrate out of psychiatry and into "real" medicine (syphilis is one well-known historical example).

Psychiatry (as well as psychopathology) is dominated by the current versions of the two diagnostic manuals, the DSM and ICD (American Psychiatric Association 2013; World Health Organization 1992).[3] In their current versions, both of these manuals are based on operational definitions, which I criticised at length in Chap. 6. It is further important to be clear about the following: The diagnostic manuals are sometimes described as being "phenomenologically descriptive" or similar (Andreasen 2007; Webb et al. 1981). This is, however, quite far from the sense of 'phenomenology' as described in the continental philosophical tradition following Edmund Husserl. Whereas the latter version of phenomenology is focused on investigating the structures of subjective experience, the standard sense of 'phenomenological description' in psychiatry is something like "a description of how symptoms appear to a neutral observer", which is, in fact, more akin to behaviourism than to continental philosophy (Parnas and Zahavi 2002, p. 139).

One consequence of the dominance of the medical model in combination with the descriptive phenomenological approach is that the study of subjective aspects of mental disorders is largely neglected.[4]

Research within psychiatry and psychopathology involves taking implicit or explicit stands on philosophically controversial issues, such as theories of mind and philosophy of science, as well as all kinds of conceptual, ethical, metaphysical, social, political, and epistemological topics. This includes difficult issues such as how best to classify mental disorders, behaviour, perceptions, sensations, and emotions. Though these issues may not necessarily pose manifest practical problems during everyday clinical activities, they certainly have important implications for the handling and treatment of psychiatric patients.

Psychopathology is an extraordinarily interesting area of research for the application of approach-based analysis of interdisciplinarity.

First, because there are straightforward arguments for why contributions from several different disciplines are needed to encompass the phenomena in question. One can hardly deny that biological, neurological as well as psychological, linguistic, and socially-orientated approaches are relevant each in their own way. Pedagogical and educational perspectives are also relevant, especially in relation to child psychiatry (Hvidtfeldt 2016).

Second, compared to interdisciplinary integrations of distant approaches such as literature studies and evolutionary theory, practitioners in psychiatry are operating with quite tangible criteria of success. The aim is to make better distinctions between various kinds of dysfunctions in ways that will eventually lead to more precise prognoses and more effective therapeutic interventions.

So even though one frequently encounters complaints that the phenomena dealt with in psychiatry are somewhat obscure and intangible, this is only partly true (e.g. in comparison to certain exemplary cases in somatic medicine, such as broken bones or heart failure). Psychiatry is in a much better position than literature studies with regard to figuring out whether some study is on the right track. Indeed, it is much easier to argue that there is a right track at all. When faced with the reality of clinical psychiatry, one must be exceedingly stubborn to deny the obvious difference between, for instance, people who are well-functioning and people who are deeply psychotic. Providing effective treatment for deeply psychotic patients would constitute a quite tangible "right track".

A further complicating factor, however, is that in psychiatry the targets are what Ian Hacking has called 'interactive kinds' (Hacking 1999), that is, kinds (of people) with the capacity to alter their behaviour as a consequence of being categorised. This means, for instance, that patients may conform to the symptomatology described in the diagnostic manuals.[5] Consequently, a considerable part of the objects studied in psychiatry are moving targets.

What is Schizophrenia?

Schizophrenia has been considered a distinct disease for more than a century. This in spite of the fact that its name, definition, and categorial boundaries have changed many times during this period, and that its aetiology and pathophysiology remain elusive and strongly contested. It is widely accepted that there is some genetic component to schizophrenia. But the nature of this component is far from clear. There is at the moment little hope of establishing genetically based diagnostics of schizophrenia.

Schizophrenia is the closest one gets to prototypical madness. It involves hallucinations of all kinds (including tactile, auditory, visual, olfactory and gustatory "visions"), delusions, paranoia, and strange (at times scary and even dangerous) behaviour. The clinical manifestations of schizophrenia are immensely diverse, however, with this heterogeneity being poorly explained by current diagnostics. Importantly, there are substantial overlaps between a number of different diagnoses in the psychosis spectrum, as well as marked heterogeneity internal to the group diagnosed with schizophrenia. In the development of DSM-5, it has been acknowledged that the heterogeneity of schizophrenia is poorly explained by the selection of subtypes of schizophrenia in DSM-IV. The treatment prescribed for this has been to simply stop subdividing schizophrenia and go for one big mess instead (Tandon et al. 2013).

Patients suffering from schizophrenia have a significantly reduced life expectancy. On top of high suicide rates and adverse side-effects of medication, schizophrenia is often correlated with socioeconomic and lifestyle risk factors such as poor diet, inadequate exercise, obesity, and smoking. All these factors translate into patients diagnosed with schizophrenia dying 12–15 years before the average population (Saha et al. 2007). This means that schizophrenia shortens lifespans more than most somatic illnesses.

The established construal of schizophrenia involves three dimensions: positive symptoms, negative symptoms, and disorganisation symptoms.

Positive symptoms, often referred to as Schneiderian first-rank symptoms (after the German psychiatrist Kurt Schneider), include delusions, hallucinations, and paranoia. They are called 'positive' since the phenomena referred to are considered to add something out of the ordinary or to be an excess of something in a clinically significant way. At the outset, positive symptoms respond well to medication.

Negative symptoms, on the other hand, are deficits of "normal" behaviour, cognition, or response. This group of symptoms includes flat expression, minimal emotional response, poverty of speech, anhedonia, lack of motivation, and lack of interest in social intercourse. Negative symptoms respond less well to medication.

The distinction between positive and negative symptoms is not all that clear, however. Though the positive symptoms of schizophrenia involve extraordinary experiences, for example, delusions of one's mind or body being controlled by some external force or agency, they also imply the absence of something which is normally present such as the sense of ownership or deliberate control (Sass and Parnas 2003, p. 431). One might further ask whether disorganisation implies the presence of a mess of the absence of order.

The following are the official diagnostic criteria for schizophrenia according to the most recent edition of DSM:

A. Two (or more) of the following, each present for a significant portion of time during a 1-month period (or less if successfully treated). At least one of these must be (1), (2), or (3):

(1) Delusions.
(2) Hallucinations.
(3) Disorganized speech (e.g. frequent derailment or incoherence).
(4) Grossly disorganized or catatonic behavior.
(5) Negative symptoms (i.e. diminished emotional expression or avolition).

B. For a significant portion of the time since the onset of the disturbance, level of functioning in one or more major areas, such as work, interpersonal relations, or self-care, is markedly below the level achieved prior to the onset (or when the onset is in childhood or adolescence, there is failure to achieve expected level of interpersonal, academic, or occupational functioning).

C. Continuous signs of the disturbance persist for at least 6 months. This 6-month period must include at least 1 month of symptoms (or less if successfully treated) that meet Criterion A (i.e. active-phase symptoms) and may include periods of prodromal or residual symptoms. During these prodromal or residual periods, the signs of the disturbance may be manifested by only negative symptoms or by two or more symptoms listed in Criterion A present in an attenuated form (e.g. odd beliefs, unusual perceptual experiences).

D. Schizoaffective disorder and depressive or bipolar disorder with psychotic features have been ruled out because either (1) no major depressive or manic episodes have occurred concurrently with the active-phase symptoms, or (2) if mood episodes have occurred during active-phase symptoms, they have been present for a minority of the total duration of the active and residual periods of the illness.

E. The disturbance is not attributable to the physiological effects of a substance (e.g. a drug of abuse, a medication) or another medical condition.

F. If there is a history of autism spectrum disorder or a communication disorder of childhood onset, the additional diagnosis of schizophrenia is made only if prominent delusions or hallucinations, in addition to the other required symptoms of schizophrenia, are also present for at least 1 month (or less if successfully treated). (American Psychiatric Association 2013, p. 99)

It should be clear that an operational definition along the above lines would include people with very divergent symptom clusters. Indeed, several patients might be diagnosed with schizophrenia without sharing a single symptom. Against that background, it is not surprising that schizophrenia is a severely heterogeneous and poorly demarcated category.

The symptoms described above all appear to presuppose an objective observer able to discriminate whether the patient lives up to certain norms or is, in fact, out of touch with reality. From a naïve point of view, experiencing a hallucination seems to rule out the possibility of being aware that what you are experiencing is not real or has no objective cause. Similarly, delusions involve false beliefs, and it seems contradictory to believe that your own beliefs are wrong.

In reality, however, matters are not that simple. Patients suffering from schizophrenia often maintain a "double bookkeeping" (Sass 2014). This means that they, on the one hand, may be convinced that a certain nurse wants to poison them, but, on the other hand, are happy to eat the food the nurse provides (despite having no death wish). Or they may be convinced that aliens drink their blood while they sleep, but readily admit that this is, of course, nonsensical.[6]

What is EASE, Then?

EASE is a research project which attempts to encompass psychiatric as well as psychopathological aspects of schizophrenia. The main people behind the project are Josef Parnas (professor of psychiatry and co-founder of the Centre for Subjectivity Research at the University of Copenhagen), Dan Zahavi (professor of philosophy and co-founder of the Centre for Subjectivity Research), and Louis Sass (Professor of Clinical Psychology at Rutgers University, New Jersey). A number of clinical and research psychiatrists and psychiatric, psychological, and philosophical PhD students and postdoctoral researchers are and have been affiliated with the project.[7]

To some extent, the developers of EASE have internalised the sociologically-orientated approach to evaluating interdiscipinarity. It is explicitly stated in various places that the project is interdisciplinary, since it

> [...] integrates recent psychiatric research and European phenomenologi-cal psychiatry with some current work in cognitive science and phenome-nological philosophy. (Sass and Parnas 2003)

This is, of course, far from a satisfying level of specificity by the standards developed in the above chapters. "Some current work in cognitive sci-ence" could hardly be less specific. I will engage further with this lack of specificity below.

Further, it is stated that "The EASE construction [...] involved senior inter-disciplinary scholars from 3 European countries" (Nordgaard and Parnas 2014, p. 1301). There is no doubt that a number of the involved researchers have impressive track records. But as argued above, this tells us little, if anything, about the epistemic merits of the integrated approach they produce. Further, their geographic locations seem irrelevant to an epistemic assessment.

The basic aim of the EASE-project is to revitalise psychopathology and psychiatry by an injection of philosophical phenomenology. More spe-cifically, the goal is to add nuance to the standard construal of schizo-phrenia by pointing out some structural changes of subjective experience which, allegedly, correlate with schizophrenia spectrum disorders. In this way, EASE is at the outset a de-idealising project in the sense that it aims to reinstate nuances which have (more or less deliberately) been idealised away within contemporary psychiatry.

It does not detract from this de-idealisation that it is a case of re-de-idealisation in the sense that psychopathology and psychiatry have previously been less idealised in this particular way. Indeed, phenomenology has been considered a vital part of psychopathology by very influential theorists in the twentieth century. Most notable, perhaps, is Karl Jaspers' efforts to develop a "Phänomenologische Forschungsrichtung in der Psychopathologie" (Jaspers 1912) (translated into English as: "The phenomenological approach in psychopathology" [1968]). Jaspers' main contributions were published quite early in the history of phenomenology—written before central works by Husserl, Heidegger, and Merlau-Ponty were available (Berrios 1992; Sass 2013, p. 96). Consequently, a new attempt at integrating psychopathology and phenomenology using updated phenomenological elements could make good sense.

Indeed, EASE differs considerably from Jaspers' phenomenological psychopathology. First, Jaspers considered schizophrenia to be fundamentally closed to empathy, thereby placing it outside the domain of reasonable targets for phenomenological inquiry (Jaspers 1963, p. 447). It is not entirely clear exactly what Jaspers meant by 'empathy', though. As he stated:

Subjective symptoms cannot be perceived by the sense-organs, but have to be grasped by transferring oneself, so to say, into the other individual's psyche; that is, by empathy. They can only become an inner reality for the observer by his participating in the other person's experiences, not by any intellectual effort. (Jaspers 1963, p. 1313)

Second, Jaspers stated that a psychiatrist

[…] can share the patient's experiences—provided this happens spontaneously without his having to take thought over it. In this way he can gain an essentially personal, indefinable and direct understanding, which, however, remains for him a matter of pure experience, not of explicit knowledge. (ibid., p. 1315)

In contrast, EASE targets subjective experiences related to schizophrenia directly from an objective perspective and certainly attempts to develop explicit, quantifiable knowledge thereof. What is also new, is that with EASE, phenomenological analysis is turned into a readily applicable tool.

An important part of the motivation for developing EASE is that current psychiatry is plagued by:

> [...] a systematic underemphasizing of the patient's subjective experience [...]. In fact, no account of human subjectivity and intersubjectivity is to be found in the contemporary psychiatric manuals, not even in the textbooks specifically dedicated to the nature of psychiatric interviewing. (Parnas and Zahavi 2002, p. 140)

EASE is intended to fill this cavity by means of psychopathological phenomenological analysis, the aim of which is "to disclose the essential, invariant properties of abnormal phenomena" (Parnas and Zahavi 2002, p. 157).

In EASE, it is suggested that closer attention should be paid to the group of negative symptoms:

> Until recently, Anglo-American psychiatry had made little effort to explore the subjective dimension of the so-called negative symptoms. (Sass and Parnas 2003, p. 433)

It is hypothesised that

> [...] the subjective experience of patients with so-called negative symptoms may involve "positive" phenomena that differ sharply from what the overt behavioral lack seems to suggest. (Sass and Parnas 2003, p. 431)

Central to EASE, then, is the attempt to capture the hidden "positive" phenomena assumed to generate the well-established negative symptoms. An important tool for this purpose is a method for conducting semi-structured interviews consisting of five more or less overlapping domains, which collectively cover 57 items related to disturbances of the so-called basic self.

To give an impression of the content of this method, the five domains and their items[8] are (Parnas et al. 2005):

1. **Cognition and stream of consciousness**

    1.1. Thought interference
    1.2. Loss of thought ipseity ('Gedankenenteignung')
    1.3. Thought pressure
    1.4. Thought block

    1.4.1. Subtype 1: blocking
    1.4.2. Subtype 2: fading
    1.4.3. Subtype 3: combination

    1.5. Silent thought echo
    1.6. Ruminations-obsessions

    1.6.1. Subtype 1: pure rumination
    1.6.2. Subtype 2: secondary rumination
    1.6.3. Subtype 3: true obsessions
    1.6.4. Subtype 4: pseudo-obsessions
    1.6.5. Subtype 5: rituals/compulsions

    1.7. Perceptualization of inner speech or thought

    1.7.1. Subtype 1: internalized
    1.7.2. Subtype 2: equivalents
    1.7.3. Subtype 3: internal as first-rank symptom
    1.7.4. Subtype 4: external

    1.8. Spatialization of experience
    1.9. Ambivalence
    1.10. Inability to discriminate modalities of intentionality
    1.11. Disturbance of thought initiative/intentionality
    1.12. Attentional disturbances

    1.12.1. Subtype 1: captivation by details
    1.12.2. Subtype 2: inability to split attention

    1.13. Disorder of short-term memory
    1.14. Disturbance of time experience

1.14.1. Subtype 1: disturbance in subjective time
1.14.2. Subtype 2: disturbance in the existential time (temporality)

1.15. Discontinuous awareness of own action
1.16. Discordance between expression and expressed
1.17. Disturbance of expressive language function

2. **Self-awareness and presence**

2.1. Diminished sense of basic self

2.1.1. Subtype 1: early in life
2.1.2. Subtype 2: from adolescence

2.2. Distorted first-person perspective

2.2.1. Subtype 1: mineness/subjecthood
2.2.2. Subtype 2: experiential distance
2.2.3. Subtype 3: spatialization of self

2.3. Psychic depersonalization (self-alienation)

2.3.1. Subtype 1: melancholiform depersonalization
2.3.2. Subtype 2: unspecified depersonalization

2.4. Diminished presence

2.4.1. Subtype 1: not being affected
2.4.2. Subtype 2: distance to the world
2.4.3. Subtype 3: as subtype 2 plus derealization

2.5. Derealization

2.5.1. Subtype 1: fluid global derealization
2.5.2. Subtype 2: intrusive derealization

2.6. Hyperreflectivity; increased reflectivity
2.7. I-split ('Ich-Spaltung')

2.7.1. Subtype 1: I-split suspected
2.7.2. Subtype 2: 'as if' experience
2.7.3. Subtype 3: concrete spatialized experience

2.7.4. Subtype 4: delusional elaboration

2.8. Dissociative depersonalization

    2.8.1. Subtype 1: 'as if' phenomenon
    2.8.2. Subtype 2: dissociative visual hallucination

2.9. Identity confusion
2.10. Sense of change in relation to chronological age
2.11. Sense of change in relation to gender

    2.11.1. Subtype 1: occasional fear of being homosexual
    2.11.2. Subtype 2: a feeling as if being of the opposite sex

2.12. Loss of common sense/perplexity/lack of natural evidence
2.13. Anxiety

    2.13.1. Subtype 1: panic attacks with autonomous symptoms
    2.13.2. Subtype 2: psychic-mental anxiety
    2.13.3. Subtype 3: phobic anxiety
    2.13.4. Subtype 4: social anxiety
    2.13.5. Subtype 5: diffuse, free-floating pervasive anxiety
    2.13.6. Subtype 6: paranoid anxiety

2.14. Ontological anxiety
2.15. Diminished transparency of consciousness
2.16. Diminished initiative
2.17. Hypohedonia
2.18. Diminished vitality

    2.18.1. Subtype 1: state-like
    2.18.2. Subtype 2: trait-like

3. **Bodily experiences**

3.1. Morphological change

    3.1.1. Subtype 1: sensation of change
    3.1.2. Subtype 2: perception of change

3.2. Mirror-related phenomena

3.2.1. Subtype 1: search for change
3.2.2. Subtype 2: perception of change
3.2.3. Subtype 3: other phenomena

3.3. Somatic depersonalization (bodily estrangement)
3.4. Psychophysical misfit and psychophysical split
3.5. Bodily disintegration
3.6. Spatialization (objectification) of bodily experiences
3.7. Cenesthetic experiences
3.8. Motor disturbances

3.8.1. Subtype 1: pseudo-movements of the body
3.8.2. Subtype 2: motor interference
3.8.3. Subtype 3: motor blocking
3.8.4. Subtype 4: sense of motor paresis
3.8.5. Subtype 5: desautomation of movement

3.9. Mimetic experience (resonance between own movement and others' movements)

4. **Demarcation/transitivism**

4.1. Confusion with the other
4.2. Confusion with one's own specular image
4.3. Threatening bodily contact and feelings of fusion with another

4.3.1. Subtype 1: feeling unpleasant, anxiety provoking
4.3.2. Subtype 2: feeling of disappearance, annihilation

4.4. Passivity mood ('Beeinflussungsstimmung')
4.5. Other transitivistic phenomena

5. **Existential reorientation**

5.1. Primary self-reference phenomena
5.2. Feeling of centrality
5.3. Feeling as if the subject's experiential field is the only extant reality
5.4. 'As if' feelings of extraordinary creative power, extraordinary insight into hidden dimensions of reality, or extraordinary insight into own mind or the mind of others

5.5. 'As if' feeling that the experienced world is not truly real, existing, as if it was only somehow apparent, illusory or deceptive

5.6. Magical ideas linked to the subject's way of experiencing

5.7. Existential or intellectual change

5.8. Solipsistic grandiosity

Most of these items are closely related to the established symptomatology
of schizophrenia. One might specifically detect the influence of phenomenology in items such as 1.10 ("Inability to discriminate modalities of
intentionality"), 1.14 ("Disturbance of time experience"), and 2.2
("Distorted first-person perspective") among others.

The EASE procedure, then, is to go through all of these items in a
semi-structured interview. That is, one is not required to start from the
top and go through all of the items one by one. Rather, one should
attempt to have a semi-natural conversation with the patient, in which all
the above issues are (eventually) covered. The ideal is a mutual, interactive reflection involving patient and interviewer in which the interviewer
prompts the patient to talk in her own words about experiences of the
relevant kinds (Parnas et al. 2005, p. 238).

The interviewer is to score all items on a five point Likert-type scale
defined as below:

0—Absence: *Definitely absent/never experienced.*

1—Questionably present: *Perhaps experienced, but either recollected only
at few occasions, or very dimly, during the patient's life.*

2—Present Mild: *Definitely experienced, at least three times in total (usually more frequently), but 2 at irregular occasions; the symptom does not
constitute a major problem or source of distress for the patient.*

4—Moderate: *Symptom is present either daily for extended periods of time
(e.g. at least daily in one week twice a year) or frequently but sporadically
over at least 12 months (may constitute a problem or a source of distress).*

5—Severe: *Almost constantly present (e.g. daily during recent 2 weeks); typically stressful, 5 source of suffering and dysfunction.*

Blank—Not scorable.

Usually when a Likert scale is used, the person examined is asked to
choose herself between the different scores. In EASE, the interviewer

scores the patient's experiences and the patient is not explicitly made aware of the scoring at all. Patients are not informed about the specific content or topics the interview is targeting. And they are not asked to confirm the scoring.

This is not necessarily invalidating. Indeed, it could be considered as a way of blinding the study in order to avoid situations in which patients attempt to behave in accordance with expectations, or, to the contrary, deny having had such experiences in order not to appear "insane".[9] Nevertheless, it does appear somewhat peculiar to blind participants from the topics of the interview, when the interview is about their private, phenomenological experiences.

Certainly, the epistemic authority of Likert scaling is far from impressive even in its original form. Certainly, there is not much to gain from purism with regard to the use of this tool. But all in all, the third-person quantitative assessment of subjective structures of EASE is quite far from the original philosophical phenomenological method in any of its versions (as will be discussed in more detail below).

On this backdrop, we are ready to embark on the specific analysis of the NP2014 approach.

The NP2014 Approach

The remainder of this chapter will be dedicated to an approach-based analysis of the approach of (Nordgaard and Parnas 2014). The NP2014 approach is quite complex and incorporates a number of parallel agendas. What is presented here is therefore a somewhat idealised representation of this interdisciplinary approach. I will focus on what is presented as 'Aim (1)' "To examine the specificity of Examination of Anomalous Self-Experiences (EASE) measured SD to the schizophrenia spectrum disorder in first contact inpatients" (p. 1300), since I consider this to be the primary investigation of the article. The remaining three aims will not be discussed. What will be discussed nevertheless reveals a number of interesting issues. The analysis shows that NP2014 incorporates a number of significant weaknesses which appear to arise from the attempts to combine elements from quite distant parent approaches. It also points out possible steps to ameliorate some aspects of this situation. Thus, the analysis shows that, at least in this case, the approach-based approach can deliver interesting and relevant results—even in a fairly concise format.

Parent Approaches?

One of the basic steps in approach-based analysis is to figure out (to the extent possible) what the specific parent approaches are. The next step is to determine which elements of the parent approaches are combined in the integrated approach in focus.

As discussed above: For all practical purposes, when working on an interdisciplinary project, you need a base somewhere (i.e. within some discipline). And when you are working from within some base discipline, you have to abide by its rules to a certain extent. This applies even if your interdisciplinary project suggests that certain basic assumptions of your home discipline are fundamentally flawed. As determined in this chapter, EASE is based firmly within psychiatry. Most EASE-related publications are in psychiatric journals or edited volumes related to psychiatry. This is quite reasonable, since the practical issues related to the problem addressed (how to best understand schizophrenia) is ultimately handled within psychiatry.

So, standard contemporary psychiatry is the base discipline. It is, however, difficult to identify one specific psychiatric approach as *the* psychiatric parent. In this case, though, standard contemporary psychiatry has very well established norms for carrying out research. This means that there are a number of standard tools, requirements, and assumptions which are quite firmly established across the discipline. Consequently, it may be acceptable to consider psychiatry as one generalised parent "approach" in this instance. And, indeed, in EASE the goal is to de-idealise by adding nuance which would affect most if not all approaches relying on the established understanding of schizophrenia.

What are the other involved approaches, then?

As is evident from several of the above quotations, phenomenological philosophy is highlighted as one of the most important inputs combined with the base in psychiatry. Phenomenological philosophy is, however, a quite heterogenous bundle of approaches, and it is required to narrow things down. The most well known representatives of phenomenological philosophy are Husserl, Heidegger, Sartre, and Merleau-Ponty. Each of these defended a distinct variety of phenomenology and phenomenological method, and they all reached different results (Smith 2013). Further, none of these agree with Karl Jaspers' take on a psychopathological phenomenology.

In NP2014, few of the elements explicitly discussed appear to be imported from classical phenomenological approaches. The influence is obvious, however, especially on the language used in certain parts of the article. Indeed, one indication of some difficulties related to the integration is that the authors appear to have trouble integrating the different elements even at the linguistic level. Most of the elements drawn from (or rather inspired by) phenomenology are described in quite rich and metaphorical language in the section entitled "Introduction". As an example, it is stated that in normal functioning cognition, "I am always already aware of "I-me-myself", with no need for introspection or reflection to assure myself of being myself" (p. 1301). In the following section, entitled "Methods", language and focus changes considerably. In this section, we hear of samples, inclusion criteria, exclusion criteria, informed consent, polydiagnostic checklists, Likert scales, Cronbach's alpha, ANOVA, and similar. None of these are discussed in any classical phenomenological writings I have come across.

We need to look at earlier EASE-related publications to figure out which elements are imported from phenomenology.

One example is Husserl's distinction between noematic and noetic aspects of consciousness:

> Noema are the intentional objects, that which consciousness is directed against, whereas noetic refers to subjective processes or structures that shape consciousness. The noetic structures in focus include distinct *modes* of intentionality (e.g., you may perceive, remember, or imagine the same object. Perceiving, remembering, or imagining are different modes of intentionality) as well as certain structures of ongoing self-awareness and embodiment the distortion of which are assumed to be fundamental in schizophrenia. (Sass and Parnas 2003, p. 429 f.)

The inspiration from these concepts is clear in some of the items of the EASE interview guide (especially 1.10 and 1.11). Another example is the assumption that through phenomenological reduction we are able to reach the (necessary and universal) core of what phenomena are (as they appear to the "reductionist"). "[P]henomenological reduction [...] is a specific kind of reflection enabling our access to the structures of subjectivity" (Parnas and Zahavi 2002, p. 158).

In EASE, and NP2014, these elements are somewhat distorted, though, given that in EASE the target is not the structure of one's own subjectivity nor, importantly, for that matter universal or necessary aspects of subjectivity. It cannot be, since by considering the subjectivity of schizophrenia patients as fundamentally disturbed their subjectivity is considered to be operating differently from the "normal" way. Structural elements or dysfunctions which are not part of well-functioning subjectivity can neither be universal nor necessary, of course. The targets in EASE are derived disturbances of structures which, through phenomenological analysis, are concluded to be necessary for the healthy operation of subjectivity. The disturbances themselves are out of reach for the phenomenological investigator (unless the investigator suffers from self-disturbances which none of the central members of the EASE team apparently do).

The targets in EASE are thus made up from phenomena which the reductionist does not have first-person access to. Phenomenological reduction is transformed into a method for accessing the core structures of subjectivity of psychiatric patients. Reaching these core structures is attempted through the interpretation of texts and verbal expressions by schizophrenic patients.

This immediately adds another complication. Noting that detailed first-person descriptions of negative symptoms are extremely rare, the French surrealist dramatist and poet Antonin Artaud is invoked as "virtually the only person with schizophrenia who has described negative symptoms in real detail". (Sass and Parnas 2003, p. 435).

There are some significant problems related to using the Artaud case. Due to his considerable success as an actor, playwright, and poet, he was certainly not your average schizophrenia patient. That he was an opium addict since the age of 22 and that some of his writings were produced in states of abstinence cast serious doubts on their status as descriptions of typical experiences of someone suffering from schizophrenia. Further, it is at least somewhat ironic to include a case such as Artaud's in historical accounts of self-disturbances, when substance abuse is an exclusion criterion for participating in EASE studies (as discussed below) as well as for being diagnosed with schizophrenia (American Psychiatric Association 2013, p. 99).

It is not unfair to say that in this version, what remains of original phenomenological approaches is transformed almost beyond recognition and put to use in ways quite different from those originally intended. There are reasons to doubt whether using this distorted version of phenomenology is suitable for the purposes of EASE.

How Distant are the Parent Approaches?

Contemporary operationalised psychiatry and philosophical phenomenology do not appear to be proximate. There are few, if any, overlaps in terms of tools and assumptions between the two disciplines. Most significantly, current psychiatry is very much a quantitative, "objective" enterprise, whereas the phenomenological focus on qualitative, subjective states and structures pulls in a quite different direction.

Jaspers' phenomenological psychopathology could perhaps be considered a middle ground between the two. It is, obviously, more closely related to psychiatry, and since this is a central source for EASE elements and assumptions (as has been specified in Parnas et al. 2013) the parent approaches may be more proximate than initial impressions admit. However, it is worth remembering that the phenomenological as well as the psychiatric elements of Jaspers' hybrid approach were very different from current psychiatry and phenomenology. Further, it is worth bearing in mind that Jaspers encountered significant difficulties with his attempt to develop a phenomenological psychopathology, and that some very central problems were never solved (Fulford et al. 2006, chap. 9).

Overall, when proximity is "measured" by the number of shared elements and assumptions, contemporary psychiatry and phenomenology (in any version) are quite distant.

The Vehicle of the Integrated Approach

What is the vehicle of representation in NP2014 and how is it related to the other approaches of EASE? The NP2014 article does not present a novel vehicle of representation. The specific NP2014 approach is constructed to test the specificity of disturbances of the basic self to schizophrenia spectrum disorders (Nordgaard and Parnas 2014, p. 1300). Thus, what is presented in NP2014 are further reasons in favour of a previously presented vehicle. The vehicle of representation of NP2014 is therefore the same one used in other EASE publications (this appears to be quite stable). We have to look elsewhere for clear expressions of what this vehicle is: Across the EASE framework, schizophrenia is represented as an *ipseity disturbance*—a disturbance of the so-called *basic self.*

> Schizophrenia, we argue, is fundamentally a self-disorder or ipseity disturbance (ipse is Latin for "self" or "itself") that is characterized by complementary distortions of the act of awareness: hyperreflexivity and diminished self-affection. (Sass and Parnas 2003, p. 427)

In the concluding section of the same article, this is repeated in slightly different terms:

> We have argued that schizophrenia is best understood as a particular kind of disorder of consciousness and self-experience. We described specific alterations of self-experience and the self-world relationship that we see as fundamental to the illness, especially diminished self-affection, hyperreflexivity, and related disruptions of the field of awareness. (Ibid., p. 439)

This, I believe, quite clearly defines a propositional model of what is believed to be fundamentally wrong with people suffering from schizophrenia. On a very soft version of the medical model, this might even be considered an acceptable explanation for the illness (though not necessarily a good one, of course).

But how, more specifically, are self-disturbances construed? Referring to Zahavi, 2005, it is stated that:

Phenomenology and neuroscience operate here with the notions of "minimal" or "core" self to describe a structure of experience that necessarily must be in place in order for the experience to be subjective, ie, to be someone's experience. (Nordgaard and Parnas 2014, p. 1301)

Throughout the EASE texts the expressions 'core self', 'minimal self', 'basic self', and 'ipseity' appear to be used synonymously (Nordgaard and Parnas 2014; Parnas 2000; Parnas et al. 1998, 2005; Raballo and Parnas 2011; Sass and Parnas 2003). It is this core, pre-reflective self which is assumed to be disturbed in schizophrenia. That assumption implicates another assumption, namely that normal functioning lives up to the ideal of an undisturbed basic self. At least this appears to be part of the background for stating that:

The minimal sense of self is always coupled with an automatic, unreflected immersion in the shared social world, variously designated, eg, "common sense," "sense of reality," "fonction du réel." The world is always there, tacitly grasped as a real and self-evident background of all experience and meaning. (Nordgaard and Parnas 2014, p. 1301)

It is even more clearly expressed in the following:

Tacitness is crucial; it is [...] the pre-reflective subjecthood, ipseity, or self-awareness that is in turn the medium through which all intentional activity is realized. Any disturbance of this tacit-focal structure, or of the ipseity it implies, is likely to have subtle but broadly reverberating effects; such disturbances must necessarily upset the balance and shake the foundations of both self and world. (Parnas 2000, p. 122)

These are quite radical statement of an (a priori) assumption about the nature of normal cognition. Can we really rule out that well-functioning individuals sometimes have slips of "self-cognisance"? Do we know that this "automatic, unreflected immersion in the shared social world" is always fully functional? Would we (through phenomenological analysis) be able to tell whether our own sense of self is sometimes disturbed or confused? Is such a generalisation reasonable?

The assumption that self-disturbances are central to schizophrenia is supported by references to a number of historical accounts in psychopathology.

> The notion of a disordered self in schizophrenia as its core phenotypic feature was articulated, in various terms and clarity in all classic texts on schizophrenia (Kraepelin, Bleuler, Minkowski, Berze, Gruhle, Jaspers, Kronfeld) [...] Kraepelin considered "disunity of consciousness" as a generative disorder in schizophrenia, whereas Eugen Bleuler listed the experiential disorders of the ego among the so-called "complex fundamental" (diagnostic) schizophrenic symptom. Jaspers observed that in schizophrenia, "Descartes' 'cogito ergo sum' (I think therefore I am) may still be superficially cogitated but it is no longer a valid experience". (Nordgaard and Parnas 2014, p. 1300)

It is unclear, though, to what extent these notions of "disunity of consciousness", "experiential disorders of the ego", or a "superficial sense of existence" are in fact similar to the notion of a disturbed "pre-reflective subjecthood". To invoke these as support for the claim that disturbances of the basic self play a significant causal role in schizophrenia (without further explication) is at least a stretch.

Schneider's discussion of a ""radical qualitative change" in the field of consciousness, comprising a disturbed first-personal perspective ("Ichheit") and a disturbed sense of "mineness" of experience ("Meinhaftigkeit")" (quoted in: Nordgaard and Parnas 2014) seems more obviously related to the EASE perspective. However, it is worth noticing that Schneider understood the phenomena he described to be generating the *first-rank symptoms* (i.e. the positive symptoms). In contrast, in EASE the disturbances of the basic self are mainly understood as causing negative symptoms and formal thought disorder (Nordgaard and Parnas 2014, p. 1301; Sass and Parnas 2003). It is, to put it in plain terms, not entirely clear to what extent these examples from the history of psychopathology actually support the central assumptions of the EASE framework.

In NP2014 it is clearly stated that self-disturbances are considered to constitute a symptom-generating foundation of schizophrenia:

> In other words, SD should not be considered as sequelae of psychosis. Rather they seem to reflect a more fundamental and generative layer of psychopathology. (Nordgaard and Parnas 2014, p. 1305)

The claim that self-disturbances play a causal role in the generation of negative symptoms and formal thought disorder is certainly intriguing. But until some account for this causal chain is presented, it is merely a bold conjecture.

As an interesting note, according to Karl Jaspers, phenomenology can tell us nothing about genesis. As Jaspers stated in his *phenomenologische psychopathologie*:

> [...] [P]henomenology has nothing to do with the genesis of psychic phenomena. Though its practice is a prerequisite for any causal investigation it leaves genetic issues aside, and they can neither refute nor further its findings. Causal studies [...] are alien to it; yet such factual investigations have been less of a danger than those "cerebral mythologies" which have sought to interpret phenomenology and replace it by theoretical constructions of physiological and pathological cerebral processes. (Jaspers 1968)[10]

EASE is not claiming anything about what causes self-disturbances, but only about the effects of having them. What Jaspers would have thought about this analysis of causality is elusive.

In EASE and NP2014, then, the vehicle of representation is a hybrid incorporating the standard model of schizophrenia from DSM and ICD with all the established symptoms retained. Added to this is the notion of a basic self working in a particular way, which may be disturbed. Such disturbances are considered to play a causal role in the generation of a subset of the symptoms characterising schizophrenia.

The Target

Succinctly put, the general target of EASE is people suffering from schizophrenia. These are represented as having certain disturbances of the basic self which are considered to generate a series of (other) symptoms.

The special contribution from NP2014 to the EASE project is that EASE-scores are compared between groups with different psychiatric diagnoses. Comparisons are between (W1) a group of patients diagnosed with schizophrenia and other nonaffective psychosis, (W2) patients diagnosed with schizotypal personality disorder (SPD), and (W3) a group of patients with other diagnoses (e.g. bipolar disorder, major depression, anxiety disorders, OCD, and non-schizotypal personality disorders).

The NP2014 approach thus has a trident structure as illustrated in Fig. 8.1. Each spearhead is targeting a distinct group of people with different psychiatric diagnoses.

This means that (almost) identical approaches are applied to three different (though closely related) targets and the results are compared (primarily by means of ANOVA) before conclusions are drawn about the vehicle.

The greyish background fields in Fig. 8.1 are meant to illustrate how the approach starts out by using standard psychiatric elements only to

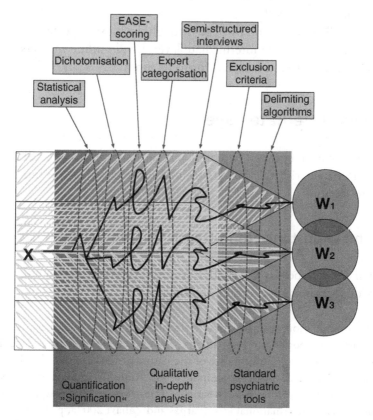

Fig. 8.1 The NP2014 approach

make a sharp shift into the use of elements picked from qualitative approaches. These are followed by a gradual return to psychiatric standards through a step-by-step quantification in order to meet the psychiatric requirements of statistically-significant results.

The main conclusion is that the application of the NP2014 approach displays a significantly higher level of self-disturbances in group W1 (schizophrenia patients) and W2 (SPD patients) compared to W3 (patients with other diagnoses).

An interesting detail is that the third spearhead, the application of the approach to W3, is included in order to falsify that the vehicle is a good representation of this particular group. The aim of this falsification is, of course, to strengthen the conviction that self-disturbances are specific to schizophrenia (and SPD).

In the following section I will go through the most important elements and assumptions of the intermediate layer. I will outline the perspective of each element and how it transforms its inputs.

The Intermediate Layer

There is a tension between two fundamental assumptions of EASE. These in turn characterise the two classes of elements which the integration aims to combine.

(1) The Importance of in-Depth Qualitative Analysis

Drawn from Jaspers is the central assumption that in-depth analyses of individual cases are of central importance:

> First, examination of single cases, as already pointed out by Jaspers, is very important. Reports from few patients, able to describe their experiences in detail, may be more informative of the nature of the disorder than big N studies performed in a crude, simplified way. Subjective experience or first-person perspective, by its very nature, cannot be averaged, except at the cost of heavy informational loss. (Parnas and Zahavi 2002a, p. 156)

Unfortunately, such assumptions are difficult to combine with well entrenched standards of medicine where the attitude towards N is "the more the merrier". Indeed, the assumption about the importance of in-depth analysis ends up in marked tension with the other most basic assumption of EASE, namely:

(2) The Significance of Quantification

The second assumption is about the importance of quantification in general and statistical significance in particular. From a standpoint within the humanities, the suggestion that phenomenology could inform psychopathology may appear interesting in itself. But in order to be acceptable to large parts of the psychiatric community, data must be quantified. It is quite difficult to publish research papers in psychiatric journals without having statistically significant results to back your conclusions (Cohen 1994; Everitt 1987). This forces the EASE team to quantify their phenomenological results and introduces strong incentives for biased representation, in case significant results are not readily achieved.

Importantly, I do not mean to suggest in any way that NP2014, or EASE for that matter, involve deliberate attempts at pretence. But as recent quite heated discussions, for instance in psychology, have shown, demands for publication involve implicit incentives to exploit "researcher's freedom" when seeking statistically significant results (so-called "p-hacking") (Simmons et al. 2011; Simonsohn et al. 2014). In some cases, p-hacking may be unintentional—in other cases researchers may believe that their goals justify the means.

The Elements

The vehicle of representation used in NP2014 is quite clearly a hybrid between current psychiatry and phenomenology in some form. But how and to what extent is the intermediate layer hybridised? The analysis hereof will show some interesting issues related to the epistemic merits of the NP2014 approach.

In the following, I will go through the most important elements of the intermediate layer. I suggest the reader keep an eye on "Fig. 8.1" above for reference.

Target Group Delimitations (Definitions/Algorithms)

Operational definitions of the DSM VI are used to pick out the target groups among first admitted patients at the Psychiatric Centre Hvidovre. Importantly, all patients were diagnosed by the authors collectively, on the basis of video interviews, notes, and hospital charts (Nordgaard and Parnas 2014, p. 1302). The importance of this aspect will be evident in several places below.

The first thing worth noticing is that the diagnostic categories of current psychiatry are retained. This choice in itself raises several interesting issues. In Chap. 1 I argued that the vehicle itself does not serve to pick out the target. One might, however, pick out one's target in ways that are more or less true to the vehicle used. In the case of EASE, the target, that is, people suffering from various mental illnesses (most importantly schizophrenia), is picked out by means of standard DSM diagnostics.

Self-disturbances may be more common in groups of people diagnosed with schizophrenia. But this does not mean that self-disturbances do not occur among people not diagnosed with schizophrenia.[11] Consequently, the extension of "people who are or will be diagnosed with schizophrenia" and the extension of "people who suffer from self-disturbances" may differ considerably. If one believes that self-disturbances should be included in the definition of schizophrenia, it appears odd to select one's target group by means of a tool which ignores these core aspects.

However, in spite of clear awareness of the problems related to operational definition in psychiatry,[12] the EASE team's hands are tied due to their base in psychiatry. In psychiatry, using official diagnostic criteria is one of the rules by which you simply have to abide in order to get your papers published as well as your projects funded.

Refusing to use official taxonomy and instead stipulating a new qualitative definition of schizophrenia (perhaps relying on continental philosophy) would amount to bibliometric suicide. When "publish or perish" rules, that strategy is out of the question. It seems safe to assume that the people involved in EASE are well aware of these issues.

Obviously, then, the targeting elements derive from psychiatry.

> *Input:*   First admitted patients at Psychiatric Centre Hvidovre.
> *Output:*   These very patients sorted in three (more or less complex) categories.
> *Issues:*   (1) All the problems discussed in Chap. 6 regarding operational definition. (2) The problem of delimiting the target without taking its (alleged) essential features into account.

Exclusion Criteria

In medical research, sets of exclusion criteria are standard tools. These are used as a means to obtain a more homogeneous target group. The exclusion criteria in NP2014 exclude patients who are aggressive, agitated, severely psychotic, or who have clinically dominating alcohol or substance abuse, a history of brain injury, mental retardation, or organic brain disorder. Involuntarily admitted and legal patients are also excluded (Nordgaard and Parnas 2014, p. 1302).

This is a pragmatic, and widely accepted, solution to a significant difficulty inherent to studies of psychiatric disorders, namely that diagnostics is entirely based on symptoms. Some patients will display symptoms simply due to drug abuse or brain damage which are not considered identical to the psychiatric disorders in question. Exclusion criteria similar to those employed in NP2014 are common-sense solutions to problems following from ignoring aetiology in diagnostics.

In NP2014, as in many other studies, the exclusion criteria also introduce substantial distortion, however. A very large proportion of first-admitted patients in the schizophrenia group (W1) are excluded on these criteria. On the other hand, a considerably smaller proportion of patients in (W2) and (W3) are excluded.

Excluding a large part of the most severely plagued patients in a study of what characterises members of a specific diagnostic category certainly raises considerable doubts as to whether conclusions drawn can be generalised to the entire category. Further, it is an empirical question whether the more "difficult" patients have higher or lower levels of self-disturbance. One could even imagine that a lower level of self-disturbance might contribute to their propensity for drug abuse or aggression. Low levels of self-disturbance would mean that they were better equipped to grasp their situation in all its difficult detail—and consequently they might be more prone to seek comfort in "self-medication".

Exclusion criteria as a tool is a psychiatric element.

Input: The three groups of patients.
Output: A reduced and homogenised version of the W1, W2, and W3.
Issues: Especially the schizophrenia group is significantly distorted since the most severely affected patients are excluded. This significantly reduces the reliability of generalisations made on the basis of the sample.

Semi-structured Interviews, Expertise, and the Likert Scale

In the following section, three distinct elements will be treated as a unit, since they are the essential components of the EASE interview method. These also constitute the central methodological contribution to the study of schizophrenia.

(1) Semi-structured Interview

Semi-structured interviews are used as a tool for gaining access to the raw data, that is, the subjective states and structures of the patients. This is a complex tool involving propositional algorithms determining the rules for how this kind of interview is carried out. There is an element of chaos[13] involved, though, in the sense that it is not possible to predict the structure a semi-structured interview will end up having. This is assumed not to be a problem though, as long as the interview ends up having covered all the items of the interview guide.

The semi-structured interview is an example of an anonymised tool imported from qualitative science. It has especially been used in areas such as sociology, anthropology, and psychology (Kvale and Brinkmann 2009, p. 130 ff.). Semi-structured interviews are often considered to provide reliable, comparable qualitative data. Indeed, as part of the EASE project the inter-rater reliability of the EASE interview guide has been assessed to be "good to excellent among trained interviewers" (Møller et al. 2011; Nordgaard and Parnas 2014, p. 1301).

In EASE, the purpose of the semi-structured interview is to contribute to bridging the gap between the qualitative focus of phenomenology and the qualitative focus of psychiatry.

It is worth noting that the particular EASE interview guide is itself hybridised by incorporating phenomenological concepts among established conceptualisations of schizophrenic symptomatology.

(2) Specialists' Categorisation

The EASE guide explicitly relies on the skills for categorisation (i.e. the ability to recognise prototypical instances of certain classes of phenomena) of "well-trained" clinicians:

> Prototype can be empirically established by examining the co-occurrence of its various features; this happens tacitly in the formation of a diagnostic skill, due to pre-reflective sedimentations of experiences and acquisition of theoretical knowledge. (Parnas and Zahavi 2002, p. 156 f.)

In this way, the categorisation skills of the "properly trained" psychiatrist are used as an instrument for measuring self-disturbances in the patients during semi-structured interviews.[14] As discussed in Chap. 6, there is a large and interesting literature in cognitive psychology focused on expert categorisation (e.g. Medin et al. 1997). Indeed, there is convincing empirical evidence that experts categorise more accurately than laymen in some contexts.

Still, there are troubling issues of calibration and bias involved. Calibration first: Ian Hacking has discussed problems of calibration related to the development of new ways of measuring intangible phenomena. Already established norms and facts may prevent one from reaching too divergent results. One of Hacking's examples was certain established convictions which IQ-tests were not allowed to challenge when originally developed. When, at first, women did better on IQ-tests than men, the tests had to be changed. It was ruled out a priori that women could be more intelligent than men, so the test required "calibration" to increase its "fidelity" to a satisfying level (Hacking 1998, p. 173). Similarly, the EASE framework involves (synthetic) a priori stipulations about which categories of people may suffer from self-disturbances. EASE tests will have to align with these framing assumptions.

A related issue is that EASE-interviewing and categorisation is liable to be affected by interviewer bias to a considerable extent. During an interview, the interviewer will be less motivated to pursue topics that indicate evidence to the contrary of what is the expected (and desired) outcome. This effect is especially strong when, as in NP2014, measurement relies on the expert categorisation of one of the persons who have diagnosed the participants in the first place (and who, as co-author of NP2014, has a strong professional interest in reaching clear and interesting results).

(3) The Likert Scale

The most significant gap to bridge, then, is that between the qualitative data gained from the phenomenologically-inspired interviews and the quantitative standards of psychiatry. It is in the effort to make these odd ends meet that the most troubling issues arise.

Qualitative data addressed in the semi-structured interviews are, via expert categorisation, quantified by means of a five-point Likert scale. The Likert scale was originally developed to measure attitudes by asking people to state the extent to which they agree with a series of statements about a given topic (Likert 1932). In an original Likert-type test, respondents are thus offered the choice between a number of fixed responses. The possible responses in EASE were noted above.

Importantly, as mentioned above, in EASE it is not the interviewee who gets to choose the answer, but the interviewer who decides whether some particular aspect of self-disturbance is present.

Using semi-structured interviews in this way, that is, in connection with assessments of subjective structures, is quite unusual in psychiatry. Perhaps this is why other standards normally taken for granted are apparently overridden. Blinding to the extent possible is considered highly desirable throughout medical science. Especially in observational studies, blinding is recommendable in order to reduce suspicions, as well as the impact, of bias. It is interesting that these basically sound elements of standard psychiatric research are "forgotten" in this interdisciplinary context.

This triad constitutes the way the subjective, experiences of the patients are registered and transformed into objective, quantitative data.

Input: The qualitative responses of the patients in the reduced and homogenised versions of W1, W2, and W3.

Output: A quantitative measure of the disturbances of subjectivity of the patients in each group.

Issues: How reliable are specialist categorisation? Especially: How reliable are the categorisations of unblinded specialists' categorisations when they have a professional interest in obtaining particular significant results?

Dichotomisation

Probably no one would consider the Likert scale to be a terribly exact instrument in any of its versions. However, it certainly does not constitute an improvement that in NP2014 the five point responses are idealised into dichotomous results. Even though one does gain something in terms of simplicity, one pays a costly price in the currency of fidelity.

The dichotomisation is achieved by means of a very simple algorithm. All items scoring "0" or "1" are subsumed as "absent". All items scoring "2", "4", or "5" are subsumed as "present". In spite of its simplicity, this manoeuvre is central and therefore deserves a brief discussion of its own.

There can be little discussion that this dichotomisation is dubious. In one blow, it removes a very significant part of the quantitative nuance supposedly obtained. It is obvious that very different results on the five-point scale may end up identical in the dichotomised version. Just think of the most extreme situation in which all patients in one group are scored as "0" (absence) on all items while all patients in a second group are scored as "5" (severe) on all items. Compare this to the situation in which all patients in group one are scored as "1" (questionably present) on all items and the patients in group two are scored as "2" (present mild) on all items. These two radically different results would appear identical after being dichotomised. Importantly, in all but the first mentioned extreme scenario the effect in the dichotomised version would be stronger than the original result.

Even more seriously, it is apparently the dichotomised figures which are fed into the statistical calculations upon which conclusions are drawn. Thus, ANOVA is used to analyse variance after most of the "measured" variance has been idealised away. I use the word 'apparently' since the data and calculations are not included in the article (and are not, as far as I can see, made publicly available elsewhere).

Seemingly, the dichotomisation algorithm is invented for the occasion. Obviously not dichotomisation itself, but rather the way it is used. There is no explanation for why this is done. One might thus get the impression that this might be a necessary move to get the statistical apparatus to produce the desired (significant) results.

The fairly rich qualitative nuances of the original data from the semi-structured interviews are increasingly reduced through these central steps of the approach. The end result is a sub-representation so radically idealised that fidelity is reduced to an absolute minimum. With no discussion of the details of these idealisations, no mentioning of the related trade-offs, and consequently no attempts to justify their use, these idealisations appear suspect at best.

> *Input:*   The quantitative five-point Likert-scale data.
> *Output:*   A dichotomous version hereof.
> *Issues:*   This manoeuvre significantly distorts the results of the central measuring method. A significant amount of nuance is removed in one blow.

Statistical Tools

The dichotomised figures are fed into the usual statistical apparatus:

> The analytic strategy was straightforward: in exploring the diagnostic distribution of the EASE scores, the diagnostic groups served as independent variable whereas the EASE scores constituted dependent variables, explored by ANOVA with polynomial (post hoc tests) analysis exploring between-group differences. (Nordgaard and Parnas 2014, p. 1303)

It is peculiar to explore the variance between three groups of patients *after* having eliminated most of the nuances of the measured variance. The analysis of the variance between dichotomised quantitative representations of a rich and complex qualitative phenomenon is certainly very far from capturing actual variation in the structures of subjectivity of psychiatric patients.

Against the background of the discussion of problems related to ANOVA above, the use of this tool should ring a bell (if not an alarm-clock). Especially since the NP2014 does not provide any information about issues deemed crucial above. Nowhere in the article are we provided with the data or the calculations related to this exploration. We see no power calculations or estimations of effect sizes. We are left only with the conclusion that deviation from the mean is significantly higher in W1 and

W2 compared to W3. Further, if the mean is the non-disturbed subjective state of healthy individuals (determined a priori?), this in itself is a rather dubious reference point.

> *Input*:   The dichotomous data.
>
> *Output*:   A significant difference between the results obtained in W1 and W2 compared to W3. These appear to corroborate that the vehicle of representation captures central aspects of the target.
>
> *Issues*:   No further comments.

The Vehicle

The output from the statistical calculation fits one aspect of the vehicle of representation neatly, in that it displays a correlation between schizophrenia (and SPD) and self-disturbance as measured via the EASE method.

The approach does not show anything about the causal role of self-disturbances in relation to negative symptoms and formal thought disorder. Neither does it indicate that self-disturbances are fundamental to schizophrenia.

The Verdict

One can hardly deny that there is a phenomenological side to suffering from schizophrenia. Since the subjective aspects of schizophrenia are absent in current diagnostics, it seems perfectly reasonable to attempt to develop a phenomenologically enriched perspective in order to capture the phenomena of schizophrenia more adequately. The quite striking responses of patients during EASE interview sessions clearly indicate that the EASE project is indeed onto something significant.[15] At the outset, then, the de-idealising aspiration of EASE and NP2014 seems justified.

The authors of NP2014 describe their interdisciplinary enterprise as

[…] a systematic qualitative and quantitative, semistructured exploration of SD […] based on the empirical data from extensive, in-depth interviews with schizophrenia spectrum patients, a review of classic and contemporary German, French, and English language literature, and conceptual inputs from philosophy of mind and phenomenology. (Nordgaard and Parnas 2014, p. 1301)

The impression one gets from this description, however, does not correspond very well to the results of an approach-based analysis. The general claim that a phenomenologically-enriched account has been developed tells us very little about how this act of representation is carried out. Zooming in by means of approach-based analysis reveals that very little of phenomenology is actually retained once the data is fitted to the vehicle.

As discussed above, the review of European psychopathological literature does not seem entirely convincing. And though the approach does include quantitative as well as qualitative elements, it seems the influence of the latter on the final results are negligible. When processed through the elements of the hybridised intermediate layer of NP2014, the qualitative data are idealised to an extent that the remains bear little resemblance to the outputs of the "extensive, in-depth interviews".

The problems detected through an approach-based analysis of NP2014 appear to be partly the result of integrating elements which do not readily combine. In fact, forcing the quantitative standards of psychiatry onto phenomenology appears to corrupt both in NP2014. On the one hand, the qualitative data obtained through the phenomenological elements are distorted almost beyond recognition in the attempt to fit these into the psychiatric framework. On the other hand, perfectly sound elements of standard psychiatry are either distorted or simply left out of the integrated approach.

Unfortunately, then, the reasonable aspirations of de-idealisation appear to be stopped in their tracks by the strict demands of the base discipline. Some of the requirements of psychiatry (especially regarding diagnostics and quantification) are not up for negotiation if one wishes to publish results in psychiatric journals. Attempts to change these aspects would be considered "unscientific" and would be met with great resistance. It is thus the basic requirements of psychiatry which prevent the desired (and, indeed, highly desirable) increase in accuracy and nuance which was the aim of EASE in the first place.

In NP2014 (and EASE in general) the elements transferred from phenomenology are transformed and used quite differently compared to the original settings:

- First, they are used to target the subjective structures of psychiatric patients, not the person carrying out the analysis. This is a requirement for using the elements in psychiatry. Obviously, it is the patients who are of interest, not the psychiatrist. This requires a shift from the first-person perspective of the phenomenologist to the third-person perspective of the psychiatrist.
- Second, it is disturbances, not healthy operation, which are in focus. Consequently, the tools of phenomenology are no longer used to study the universal and necessary structures of subjectivity, but rather disturbances of structures of subjectivity. Does this mean that the targeted "structure of experience that necessarily must be in place in order for the experience to be subjective" is not necessary after all? At least it does not seem that patients diagnosed with schizophrenia lack subjective experience as such. One could argue that EASE is targeting minor disturbances which do not undermine subjective experience in general, but still have detrimental effects on aspects of mental health. Even in this case, though, there is no doubt that the use of phenomenology is changed considerably, and that the altered use appears less reliable than the original one.
- Third, processing the qualitative data from the phenomenological interviews through expert categorisation, quantification via Likert scale, and dichotomisation, is a process of strong idealisation which almost entirely removes the qualitative richness of the raw data.

The alterations of the qualitative and phenomenological elements in the integrated approach are quite considerable. The alterations are implemented in order to make these elements fit better to the requirements of the base discipline.

There are considerable changes in the base discipline as well, however.

In many approaches to qualitative science, there is little utility of tools such as blinding, randomisation, and control groups. That qualitative elements are involved in the NP2014 approach does not make elements such as these superfluous, however.

Not having a control group is equal to deciding a priori what is the normal level of self-disturbances. It should be clear that this is not part of a virtuous scientific method. How common is "magical thinking"[16] (item 5.6), "cenesthetic experiences" (item 3.7), or "a feeling as if being of the opposite sex" (item 2.11.2)? Controversies related to pathologising issues of gender and superstition aside, it seems unwarranted to decide a priori on the prevalence of these "disturbances" in the population at large. Until some empirically-supported understanding of how common self-disturbances are among healthy individuals is established, it is difficult to determine whether items such as those belong in definitions of mental pathologies.

Moreover, the lack of normal controls potentially threatens the fragile significant differences between the different diagnostic groups studied. In case it turned out that healthy controls have a higher level of self-disturbance than people suffering from major depression[17] the statistical significance might just vanish into thin air.

Abandoning tools for constraining bias, such as blinding in various forms, is not a healthy move either. It certainly reduces one's trust in the impartiality of the study. But the inclusion of healthy controls and double blinded study-designs are gold standards in psychiatry (and somatic medicine).[18] One may only speculate as to why these scientifically sound elements have been abandoned in NP2014. There are no discussions of (and apparently, no legitimate reasons for) this lowering of standards in NP2014 and other EASE-related approaches. Leaving out these fundamental elements is a significant neglect, which can hardly be defended. Unless, of course, one accepts the a priori phenomenological characterisation of the state of a normal cognitive system and firmly believes in the neutrality of semi-structured interviews and the process of quantification.

A further unfortunate consequence of the integration is that through the use of statistical tools and the reporting of impressive significance levels, self-disturbances appear as if they have been measured with a fair degree of certainty and accuracy. Reality is quite different: In fact, self-disturbances are strongly idealised and operationalised as score on the EASE scale. On this background, one must conclude that the approach of NP2014 is a very low fidelity approach indeed.

The output of NP2014 is that to make progress in the study of schizophrenia we should include phenomenology in our scientific and clinical approaches to schizophrenia. Intuitively, this suggestion seems reasonable. But based on an approach-based analysis, it is doubtful that this recommendation carries much weight. This interdisciplinary approach does not seem to deliver as promised.

A more reasonable conclusion might have been that *"our studies so far indicates that there are some very interesting subjective disturbances correlating (to some extent) with schizophrenia spectrum disorders. Unfortunately, fundamental disciplinary requirements and restrictions of psychiatry prevent us from studying these in the most optimal way (or at least publishing the results). Our conclusion is that psychiatric science needs to be less focused on quantifiable measures and more inclusive with regards to qualitative research methods and results in order to allow studies of these issues to move forward"*.

This is quite far from the actual output of NP2014, with its conclusive declamations that disturbances of ipseity have been measured quantitatively with a fair degree of certainty.

On closer inspection, it is clear that one cannot reasonably conclude that self-disturbances constitute a fundamental generative aspect of schizophrenia. Rather, one may conclude (far more modestly) that in an EASE test you are more likely to get a somewhat higher score if you are also diagnosed with Schizophrenia, SPD or non-affective psychosis.

We cannot conclude anything about causal relations. Indeed, it does not seem that there is much reason to make assumptions about causal relations between SD and schizophrenia, since non-schizophrenic patients score quite high on the EASE scale as well (despite the bias). Remove the dichotomisation. Remove bias. Add blinding and a control group. What effect would be left? It is impossible to tell, but perhaps nothing. On this background, it is it difficult to assess to what extent NP2014 adds nuance or confusion to the discussion of schizophrenia.

Whether EASE and NP2014 are epistemically more or less virtuous than the parent disciplines is of course *the* central question. In the perspective of approach-based analysis, the answer is "less".

The phenomenological elements deliver far more rich and nuanced descriptions in the parent disciplines than in NP2014. In EASE and NP2014 there is less direct access to the phenomena targeted, and explorations of these phenomena are less reliable. Rather than focusing on the essence of something which you can observe, focus is on disturbances of assumed essences of something which is not directly observable. The means for capturing the phenomena of interest are dubious at best, and the same goes for the quantitative sub-representations produced.

On the other hand, standard psychiatric methods are more rigorous and devoted to avoiding bias than NP2014. Avoiding bias by means of blinding is indeed a virtue usually associated with psychiatric approaches. Not even attempting to blind very central parts of the NP2014 approach is certainly less virtuous in this respect. No doubt, standard psychiatry is also plagued by p-hacking, bias, the file-drawer problem, calibration to fit established norms, and related issues. But in NP2014 some of the standard tools for handling such difficulties are removed with nothing to replace them.

The Good News

Though NP2014 involves low fidelity representation it does reveal some noteworthy issues. One thing it points out convincingly (though, perhaps, not surprisingly) is that standard psychiatry inadequately represents schizophrenia (as well as other ways of being mentally disordered). It would probably be a good idea to retain categories such as schizophrenia and schizotypal personality disorder until something better is established. In the meantime, EASE and other "deviant" psychiatric research programmes should continue pointing out aspects of mental illness which future developments of psychiatric and psychopathological theory will need to take into account.

In spite of the weaknesses of NP2014 discussed above, there are indications that approaches of the EASE project corroborate at least some of the distinctions made in the official taxonomies in the sense that a high EASE score is, after all, more common among patients diagnosed with schizophrenia and SPD and less common among patients with most other diagnoses. Especially interesting in this respect, is that high EASE scores appear to be predictive of future diagnoses of schizophrenia.

This certainly indicates that there are good reasons to continue developing the research programme. The integration of phenomenology and psychiatry may at some point develop into a fruitful enterprise. Indeed, attempts to do so may highlight weaknesses in the parent approaches which researchers would do good to address. But one should carefully consider the combined elements to avoid importing and combining bad habits such as the assumption that achieving a statistically significant result (once) demonstrates anything (Goodman 2016).

If the difficulties discussed above are addressed adequately, future more refined approaches relying on the EASE framework may contribute to increasing both dynamic and representational fidelity of scientific representations of schizophrenia. This could increase possibilities of prediction, which would constitute a significant step forward regarding early detection and prognosis (especially regarding prodromal and pre-psychotic patients). Such developments will, however, require more meticulous efforts in the construction of approaches as well as more flexibility in the base discipline.

What Causes the Problems?

The phenomenological elements of the NP2014 vehicle of representation are pale versions of the rich and nuanced qualitative representations one might have hoped for. The remnants of phenomenology in the vehicle of representation have little left of the qualitative richness of the data from the original empathic in-depth interview. This is partly a consequence of maintaining the demand for statistical significance and delimiting target groups by means of operationalised diagnostics.

Even though the problems just mentioned (as well as those related to quantification, bias, and lack of blinding) are distinctly epistemic in nature, a significant part of their causes are social. Looking at the social dimension of EASE provides more than a hint of an answer to why NP2014 and EASE are constructed as they are, namely that the EASE project is deeply embedded in psychiatry. Innovative interdisciplinary studies such as EASE are almost per definition in opposition to established standards. But the base discipline (and to some extent other disciplines involved) constrains the interdisciplinary scientists' creative and innovative freedom.

This is worth noting, since it illustrates that even though approach-based analysis highlights central epistemic aspects of interdisciplinary representation, in many cases we need to draw on the social and purpose related dimensions of the involved disciplines in order to explain why these particular elements have been chosen and combined in a particular way. In many cases, a detailed historical account is required in order to explain why a particular element plays some particular role, or how some assumption ended up as fundamental in a given discipline. The discussion of operational definition in Chap. 6 is a good example hereof.

To Do-List

The following is a list of potential adaptions that would considerably heighten the scientific quality in future updates of EASE and NP2014:

- First, blind the approach (to the extent possible). There are a lot of difficulties related to fully blinding studies which involve direct interaction with psychiatric patients. For instance, the trained clinician will often easily distinguish patients liable to receive a schizophrenia diagnosis from, say, patients with a major depression. Nevertheless, the person who interviews and scores the patients on the EASE scale should, to the extent possible, be unaware of diagnosis and other clinical information. Especially in cases where semi-structured interviews are used, the interview process is highly sensitive to bias. At the very least, the scoring on the EASE scale should not be carried out by a person with obvious professional interests in obtaining certain results.[19]
- Second, always include a group of healthy controls. Even if related approaches have shown certain characteristic differences between the

sample and healthy controls, a new (perhaps slightly altered approach) may reach different results. That there are significant differences between two groups of patients, may certainly point to something interesting. However, as discussed above, if variances were measured against a backdrop of healthy controls who scored somewhere in between the groups of diagnosed patients, significance might vanish into thin air.[20] This is all very speculative, of course. But so is the a priori decision to consider self-disturbance to be absent in healthy individuals.

- Third, the mathematics should be upgraded in order to retain more nuance. This would perhaps also hurt statistical significance, but that is hardly a well-founded reason for retaining sloppy standards. Certainly, capturing the five-point variance mathematically is much more complex than handling the dichotomised version. But fidelity is also much lower in the dichotomised version (and it wasn't too impressive in the first place).

- Fourth, claiming that some specific disturbances of basic structures of consciousness only to delimit one's target by means of operational definitions which pay no attention to these structures is odd, to say the least. It would be quite remarkable (if not miraculous) if the sample pointed out by the operational definition of schizophrenia corresponded exactly to those with the strongest self-disturbances.

When one might apparently detect these disturbances in "patients to be" even before the prodromal[21] phase of the illness (Møller and Husby 2000; Parnas et al. 1998, 2011), why not replace the expression "People with schizophrenia tend to display ipseity disturbance" with "people with ipseity disturbances tend to end up with a diagnosis of schizophrenia"? Why retain schizophrenia as the primary category? In fact, if self-disturbances are as fundamental as suggested, why not go all in and opt for defining what would be a truly revolutionary natural kind-like category of an "ipseity disturbance disorder"? One response might be that self-disturbance is only one aspect of the "coherent" clinical picture of schizophrenia. However, if indeed disturbances of the basic self are fundamental to schizophrenia, and detectable in prodromal and earlier phases, it seems reasonable to at least attempt using this as the basic distinction.

Fifth, if one wants to use literature reviews to add a scholarly dimension to one's studies, one should put in considerably more effort. The literature review of NP2014 is a strongly idealised version of the careful and thorough literary reviews we know from the humanities. The best example, since the topic is closely related, may once again be Patricia Kitcher's work on Freud (1992, 2007). Compared to such analytical efforts, the literary reviews of EASE come to appear cursory at best. As the discussion above suggests, a number of weaknesses of the discussions of the history and literature of psychopathology are quite easily pointed out. Thus, the literature review in its present form does not lend much support to the central assumptions of EASE. However, in a more thorough version, it might certainly do so.

Summing Up

There is no doubt that the interdisciplinary EASE project contains good ideas. Capturing the subjective experience of what it is like to suffer from schizophrenia would certainly enrich our best representations of schizophrenia considerably. And further, if specific disturbances to structures of consciousness could be determined, this ought to attract a lot of attention in psychopathological research and might have considerable impact on future developments in psychiatric nosology and diagnostics.

I do not mean to conclude that EASE in general (or NP2014 in particular) are tokens of bad science or bad interdisciplinarity for that matter. However, on an approach-based analysis it appears that there are a number of issues related to the integration of the involved parent approaches which are neither explicitly addressed nor easily overcome. If these issues were to be handled adequately, however, EASE in its current form may constitute a significant step on the way towards developing richer and more adequate representations of the class of sufferings currently labeled as 'schizophrenia'.

Approach-based analysis reveals that at least some of the problems derive from not paying sufficient attention to difficulties specific to the integration of distinct and rather distant approaches. These problems are particular to interdisciplinarity, I claim.

The difficulties detected through approach-based analysis may not necessarily be unsolvable. But solving them would require more meticulous care and attention to details of disciplinary integration. This said, it is rather difficult to imagine an integration of these particular (or related) parent approaches, which would not result in substantial (detrimental) distortion of some of the involved elements.

As discussed, EASE is a de-idealisation project. EASE does not purport to make bold conjectures or to leap into unknown territory leaving already established knowledge behind. To some extent this is the Achilles' heel of the project. For instance, one central problem is that the attempt at de-idealisation involuntarily imports the poor tools for diagnosis inherent to standard psychiatry. Would this have been obvious on existing approaches to analysing interdisciplinarity? I think not. Especially not if social aspects were kept in central focus.

One could certainly do a much more detailed analysis. For instance, one could dissect the interview guide in closer detail and analyse the different concepts picked from various phenomenological and psychopathological sources. This might reveal further interesting issues. However, it does not seem required in order to gain an impression of some central problems in NP2014, to point out reasonable "next steps" if one wants to improve upon the NP2014 approach, or, importantly, to illustrate the utility of approach-based analysis.

Against the above background, one cannot draw general conclusions about interdisciplinary science, of course. Perhaps the difficulties of integration are specific to the EASE project. But one may conclude that such difficulties do appear to exist, and that at least in some cases they are sufficiently debilitating to justify careful analysis along similar lines in other cases.

Indeed, even if some were to disagree with (or even take offence at) the conclusions reached about EASE and NP2014 above, I hope to still leave the impression that approach-based analysis constitutes a viable method for disentangling elements of interdisciplinary projects and for making progress towards being able to deal with the epistemic assessment thereof.

Notes

1. We need to distinguish clearly between the EASE project and the distinct constitutive approaches as presented in individual publications. To repeat myself: An approach-based analysis of interdisciplinary activities requires focus on one interdisciplinary approach (at a time). As discussed in several places above, in approach-based analysis specific research projects (or programmes, perhaps) are considered to be bundles of closely related approaches. A discipline, on the other hand, is considered to be a bundle (of bundles) of approaches. For convenience and clarity, I will use 'NP2014' to refer to the approach in focus in this case study and 'EASE' to refer the bundle, of which NP2014 is a constitutive element.

2. With the important caveat that I am not a trained psychiatrist. Consequently, I do not possess the competences to carry out any kind of psychiatric diagnostics with or without the EASE-methods involved.

3. Short for 'Diagnostic and Statistical Manual of Mental Disorders' (published the American Psychiatric Association) and 'The International Classification of Diseases' (published by WHO).

4. In Hvidtfeldt (2011), I analysed differences between how anxiety patients and psychiatrists described what it is to suffer from anxiety. Interestingly, in the patients' descriptions expressions such as 'being afraid', 'feeling scared', 'to panic, or 'feeling terrified' were absolutely central, but these played only a very minor role in descriptions of anxiety by the psychiatrists.

5. Another interesting finding of (Hvidtfeldt 2011) was that anxiety patients in some cases seemed to develop the official symptoms of anxiety only after being presented with the suspicion that anxiety might be the explanation for some indeterminable somatic complaint (such as more or less chronic bellyache) for which they had consulted their GP. Importantly, the symptoms of anxiety are perfectly healthy reactions in threatening situations. Indeed, it does not seem completely irrational to be more than a little alarmed by being diagnosed with a mental disorder.

6. For interesting philosophical discussions of delusions and double book-keeping, see Bortolotti 2010a, b; Murphy 2012.

7. It would not be unreasonable to describe EASE as mainly proceeding in the polymath mode with all the associated risks discussed in Chap. 3. Even though it is an interesting topic, I shall refrain from providing

anecdotal evidence for this claim, since it is not central to the goal of this case study.

8. All items are discussed in some detail and exemplified in (Parnas et al. 2005, pp. 240–256), to which I refer the interested reader.

9. Recall Hacking's notion of "interactive kinds" briefly mentioned above.

10. Apart from his opinion about the utility of phenomenology in studies of genesis, it is interesting that Jaspers appear to express a strong disdain for interdisciplinarity in this quote. The danger of "cerebral mythologies" can best be understood along the lines of "the seductive allure of neuroscience" discussed in Chap. 3 (Weisberg et al. 2008). Certainly, Jaspers did not endorse a strong medical model.

11. This is likely to be the case even if we ignore the significant problems of false positives and false negatives which plague psychiatric diagnostics (Cohen 1994, p. 998 f.).

12. Josef Parnas has published extensively on various problems of contemporary psychiatry, for instance those related to operationalisation (Parnas and Bovet 2014).

13. In the technical sense of a dynamical system which is highly sensitive to initial conditions. See Kellert 2009 for an interesting discussion of chaos theory and interdisciplinarity.

14. In a less friendly tone, Kieran Healy criticises reliance on the sophisticated judgment of people in possession of esoteric expertise. He claims that a "weak methodological core invites connoisseurs", and that "[…] connoisseurship thrives best in settings where judgment is needed but measurement is hard" (Healy 2017).

15. This is most apparent in live interview sessions and video recordings. For examples in print, see Parnas et al. (2005).

16. That is, "Ideas Implying Nonphysical Causality" (Parnas et al. 2005, p. 255). Examples could be the belief that hoping for something to happen will affect whether it happens or not, or that the weather is somehow affected by one's mood.

17. Certainly, it would not take a talented evolutionary psychologist very long to come up with a list of advantages related to an optimum level of self-disturbance. For instance, it might be easier to maintain a high spirit if one's somewhat disturbed view of the world would facilitate the "reinterpretation" of facts to fit one's preferences better.

18. Which the authors of NP2014 are well aware of, of course, and make use of in other contexts (Parnas et al. 2011, p. 201).

19. The importance of this point is not diminished by (EASE) studies suggesting high interrater reliability such as presented in (Møller et al. 2011; Nordgaard and Parnas 2012). High interrater reliability may suggest, but does not demonstrate neutrality (i.e. immunity towards bias). Further, unfortunately, the mentioned interrater reliability studies are not terribly convincing. Though the raters compared were blinded to diagnostic and clinical information in the studies, they did not rate on the basis of independent interviews. Actually, they were either both present during the interview or re-rated on the basis of videotaped interviews. Since a central reason for doing semi-structured interviews is the possibility of addressing interesting issues in detail, it is hardly surprising if a second rater notices the same interesting topics that the interviewer finds worth pursuing in more detail.

20. In spite of these hypothesised smaller effect sizes, one might still get significant results, of course, but that might require substantially larger samples.

21. "Prodromal" is used to capture a phase in which symptoms of an illness are experienced before the full-blown syndrome has developed. The concept implies that the development of the full-blown illness is inevitable (Phillips et al. 2005). In the case of schizophrenia, it is often negative symptoms that appear in the prodromal phase. This is in part a consequence of the requirement of at least one positive symptom for being diagnosed with schizophrenia.

References

American Psychiatric Association. 1980. *Diagnostic and Statistical Manual of Mental Disorders*. Edited by American Psychiatric Association. 3rd ed. Arlington, VA: American Psychiatric Association.

———. 1994. *Diagnostic and Statistical Manual of Mental Disorders*. 4th ed. Washington, DC: American Psychiatric Association.

———. 2000. *Diagnostic and Statistical Manual of Mental Disorders*. 4th ed. Washington, DC: American Psychiatric Association.

———. 2013. *Diagnostic and Statistical Manual of Mental Disorders*. Edited by American Psychiatric Association. 5th ed. Washington, DC: American Psychiatric Association.

Andreasen, Nancy C. 2007. DSM and the Death of Phenomenology in America: An Example of Unintended Consequences. *Schizophrenia Bulletin* 33 (1): 108–112.

Berrios, German E. 1992. Phenomenology, Psychopathology and Jaspers: A Conceptual History. *History of Psychiatry* 3 (11): 303–327.

Bortolotti, Lisa. 2010a. *Delusions and Other Irrational Beliefs, International Perspectives in Philosophy and Psychiatry*. New York: Oxford University Press.

———. 2010b. Double Bookkeeping in Delusions: Explaining the Gap between Saying and Doing. In *New Waves in the Philosophy of Action*, ed. K. Frankish, A. Buckareff, and J. Aguilar. Basingstoke, UK: Palgrave.

Cohen, Jacob. 1994. The Earth is Round (P < .05). *American Psychologist* 49 (12): 997–1003.

Everitt, Brian S. 1987. Statistics in Psychiatry. *Statistical Science* 2 (2): 107–116.

Fulford, K.W.M., et al. 2006. *Oxford Textbook of Philosophy and Psychiatry*. New York: Oxford University Press.

Goodman, Steven N. 2016. Aligning Statistical and Scientific Reasoning. *Science* 352 (6290): 1180–1181. https://doi.org/10.1126/science.aaf5406.

Hacking, Ian. 1998. *Mad Travelers: Reflections on the Reality of Transient Mental Illnesses*. Charlottesville, VA: University Press of Virginia.

———. 1999. *The Social Construction of What?* Cambridge, MA: Harvard University Press.

Healy, Kieran. 2017. Fuck Nuance. *Sociological Theory* 35 (2): 118–127.

Hvidtfeldt, Rolf. 2011. *Conceptual Plasticity—A Sober Constructivist Account of Categorization + Consequences*. Master thesis, Philosophy, University of Copenhagen.

———. 2016. Sådan får man et diagnosebarn. *VERA—tidsskrift for pædagoger* 74: 54–56.

Jaspers, Karl. 1912. Die phänomenologische Forschungsrichtung in der Psychopathologie. *Zeitschrift für die gesamte Neurologie und Psychiatrie* ix: 391–408.

———. 1963. *General Psychopathology*. Translated by J. Hoenig and M. Hamilton. Manchester: Manchester University Press.

———. 1968. The Phenomenological Approach in Psychopathology. *The British Journal of Psychiatry* 114 (516): 1313–1323.

Kellert, Stephen H. 2009. *Borrowed Knowledge: Chaos Theory and the Challange of Learning Across Desciplines*. Chicago, IL: University of Chicago Press.

Kitcher, Patricia. 1992. *Freud's Dream: A Complete Interdisciplinary Science of Mind*. Cambridge, MA; London: MIT Press.

———. 2007. Freud's Interdisciplinary Fiasco. In *The Prehistory of Cognitive Science*, ed. Andrew Brook, 230–249. Basingstoke, UK; New York: Palgrave Macmillan.

Kvale, Steinar, and Svend Brinkmann. 2009. *Interviews—Learning the Craft of Qualitative Research Interviewing*. Thousand Oaks, CA: SAGE Publications, Inc.

Likert, Rensis. 1932. A Technique for the Measurement of Attitudes. *Archives of Psychology* 140: 1–55.

Medin, Douglas L., et al. 1997. Categorization and Reasoning among Tree Experts: Do All Roads Lead to Rome? *Cognitive Psychology* 32: 49–96.

Møller, Paul, et al. 2011. Examination of Anomalous Self-experience in First-Episode Psychosis: Interrater Reliability. *Psychopathology* 44: 386–390.

Møller, Paul, and R. Husby. 2000. The Initial Prodrome in Schizophrenia: Searching for Naturalistic Core Dimensions of Experience and Behavior. *Schizophrenia Bulletin* 26: 217–232.

Murphy, Dominic. 2006. *Psychiatry in the Scientific Image.* UK: MIT Press.

———. 2012. The Folk Epistemology of Delusions Dominic Murphy. *Neuroethics* 5: 19–22. https://doi.org/10.1007/s12152-011-9125-5.

Nordgaard, Julie, and Josef Parnas. 2012. A Semi Structured, Phenomenologically-Oriented Psychiatric Interview: Descriptive Congruence in Assessing Anomalous Subjective Experience and Mental Status. *Clinical Neuropsychiatry* 9 (3): 123–128.

———. 2014. Self-disorders and the Schizophrenia Spectrum: A Study of 100 First Hospital Admissions. *Schizophrenia Bulletin* 40 (6): 1300–1307. https://doi.org/10.1093/schbul/sbt239.

Parnas, Josef. 2000. The Self and Intentionality in the Pre-Psychotic Stages of Schizophrenia. A Phenomenological Study. In *Exploring the Self*, ed. Dan Zahavi, 115–147. Amsterdam: John Benjamins Publishing Company.

Parnas, Josef, et al. 1998. Self-experience in the Prodromal Phases of Schizophrenia. *Neurology, Psychiatry and Brain Research* 6: 97–106.

———. 2005. EASE: Examination of Anomalous Self-experience. *Psychopathology* 38: 236–258.

———. 2011. Self-experience in the Early Phases of Schizophrenia: 5 Year Follow-Up of the Copenhagen Prodromal Study. *World Psychiatry* 10: 200–204.

———. 2013. Rediscovering Psychopathology: The Epistemology and Phenomenology of the Psychiatric Object. *Schizophrenia Bulletin* 39 (2): 270–277. https://doi.org/10.1093/schbul/sbs153.

Parnas, Josef, and Dan Zahavi. 2002. The Role of Phenomenology in Psychiatric Diagnosis and Classification. In *Psychiatric Diagnosis and Classification*, ed. Mario Maj, Wolfgang Gaebel, Juan Jose López-Ibor, and Norman Sartorius. Chichester, UK: John Wiley & Sons, Ltd.

Parnas, Josef, and Pierre Bovet. 2014. Psychiatry Made Easy: Operation(al)ism and Some of Its Consequences. In *Philosophical Issues in Psychiatry III: The Nature and Sources of Historical Change*, ed. Kenneth S. Kendler and Josef Parnas. Oxford, UK: Oxford University Press.

Phillips, Lisa J., et al. 2005. Prepsychotic Phase of Schizophrenia and Related Disorders: Recent Progress and Future Opportunities. *The British Journal of Psychiatry* 187 (48): 33–44. https://doi.org/10.1192/bjp.187.48.s33.

Raballo, Andrea, and Josef Parnas. 2011. The Silent Side of the Spectrum: Schizotypy and the Schizotaxic Self. *Schizophrenia Bulletin* 37 (5): 1017–1026. https://doi.org/10.1093/schbul/sbq008.

Saha, Sukanta, et al. 2007. A Systematic Review of Mortality in Schizophrenia: Is the Differential Mortality Gap Worsening Over Time? *Archives of General Psychiatry* 64 (10): 1123–1131.

Sass, Louis A. 2013. Jaspers, Phenomenology, and the 'Ontological Difference'. In *One Century of Karl Jaspers' General Psychopathology*, ed. Giovanni Stanghellini and Thomas Fuchs. Oxford: Oxford University Press.

———. 2014. Delusion and Double Book-Keeping. In *Karl Jaspers' Philosophy and Psychopathology*, ed. Thomas Fuchs, Thiemo Breyer, and Christoph Mundt, 125–148. New York: Springer.

Sass, Louis A., and Josef Parnas. 2003. Schizophrenia, Consciousness, and the Self. *Schizophrenia Bulletin* 29 (3): 427–444.

Simmons, Joseph P., et al. 2011. False-Positive Psychology: Undisclosed Flexibility in Data Collection and Analysis Allows Presenting Anything as Significant. *Psychological Science* 22: 1359–1366. https://doi.org/10.1177/0956797611417632.

Simonsohn, Uri, et al. 2014. P-Curve: A Key to the File-Drawer. *Journal of Experimental Psychology: General* 143 (2): 534–547. https://doi.org/10.1037/a0033242.

Smith, David W. 2013. Phenomenology. In *The Stanford Encyclopedia of Philosophy* (Winter 2013 Ed.), ed. Edward N. Zalta. https://plato.stanford.edu/archives/win2013/entries/phenomenology/.

Tandon, Rajiv, et al. 2013. Definition and Description of Schizophrenia in the DSM-5. *Schizophrenia Research* 150 (1): 3–10.

Webb, L., et al. 1981. *DSM-Ill Training Guide for Use with the American Psychiatric Association's Diagrostic and Statistical Manual of Mental Disorders*. 3rd ed. New York: Brunner/Mazel.

Weisberg, Deena Skolnick, et al. 2008. The Seductive Allure of Neuroscience Explanations. *Journal of Cognitive Neuroscience* 20 (3): 470–477.

World Health Organization. 1992. *The ICD-10 Classification of Mental and Behavioral Disorders*. Geneva: World Health Organization.

Zahavi, Dan. 2005. *Subjectivity and Selfhood: Investigating First Person Perspective*. Cambrigde, MA: The MIT Press.

9

Conclusion

In this final chapter, I will briefly sum up and comment on the discussions above. But before doing so, I will remind the reader of the following ancient recommendation regarding proper scientific conduct:

> It is the mark of an educated person to look for precision in each class of things just so far as the nature of the subject admits: it is evidently equally foolish to accept probable reasoning from a mathematician and to demand from a rhetorician demonstrative proofs. (Aristotle, 1094b, pp. 24–25)

In this book, I have developed a framework for epistemic assessment of interdisciplinarity in the sense of integration of scientific approaches. I have discussed epistemic assessment as the evaluation of whether a particular integrated approach is more or less virtuous than its parents when measured against more or less well-established epistemic ideals. That is, for instance, whether integrated approaches deliver increased explanatory power, additional detail or nuance, improved accuracy (e.g. in terms of prediction or distinction), increased scope, more general implications, superior conceptual coordination, improvements in terms of cognitive economy (a.k.a. simplicity), improved ability to control, produce, or prevent specific phenomena,

© The Author(s) 2018
R. Hvidtfeldt, *The Structure of Interdisciplinary Science*, New Directions in the Philosophy of Science, https://doi.org/10.1007/978-3-319-90872-4_9

improved empirical adequacy, improved dynamical and representational fidelity, increased transparency, strengthened replicability, improved reliability, or increased explicitness regarding idealisations and distortions.

Bad interdisciplinarity is bad science. A central question I have tried to address is whether there are pitfalls specific to interdisciplinary science. I believe I have provided an affirmative answer to this question above. When we disentangle an interdisciplinary approach, and determine the elements out of which it is constructed, it is clear that these have often been developed for particular uses in particular circumstances. Due to the non-universality of the elements of representation, relocating elements between contexts will involve the risk of violating ceteris paribus clauses. When we add to this that scientists constructing interdisciplinary approaches in many cases do not possess deep expertise with respect to the disciplines from which they import elements, it is likely that the background knowledge required for using each element "properly" is often lacking. I have discussed this issue under the heading of 'relocational idealisation'.

I believe approach-based analysis brings us a good deal closer to being able to draw conclusions about epistemic issues related to interdisciplinary science. It does so by revealing a number of central aspects of scientific crossbreeding which have hitherto been covered in darkness. As illustrated above, there are many complex questions regarding the epistemic value of interdisciplinary activities. Since a lot of these questions are entirely unaddressed in the central literature and debates on interdisciplinarity, approach-based analysis certainly constitutes a step forward in this respect.

The extent to which the efforts invested in the development of approach-based analysis have paid off cannot be determined conclusively on the evidence provided in this book, however. A philosophical discussion and a single case study hardly suffices in this respect.

Obviously, the argument in favour of approach-based analysis would have been stronger had I been able to present a number of different case studies, each revealing interesting additions to prima-facie impressions of their integration. This point is crucial, since a central conclusion above has been that whether interdisciplinary science result in better acts of representation must be evaluated through careful analysis on a case-by-case basis.

Though the single case study above cannot settle all the relevant issues, it does indicate that a good place to start such case-by-case analyses is by determining the vehicles used, identifying the elements of the intermediate layer, figuring out how the use of these change as part of the integration, and how all this affects various epistemic standards. In this way, this book presents good reasons for paying close attention to representation, as well as issues of distortion, when analysing interdisciplinary science. Articulating epistemic aspects of the integration of approaches as suggested above thus has the potential to significantly affect how we evaluate particular cases of interdisciplinarity.

Indeed, there are a few more reasons to be optimistic with regard to the general utility of approach-based analysis, I believe. One is that approach-based analysis has proven to be quite easily applicable in the case study in Chap. 8. Another is that the EASE case study drew out a number of interesting aspects which would not have shown on conventional accounts of interdisciplinarity. In this way, approach-based analysis appears to be useful for disentangling what it is exactly that makes, for instance, particular approaches in cognitive psychology "cognitive", approaches in neuro-aesthetics "neuro-like", and approaches in evolutionary art theory "evolutionary".

One complication, however, is that in order for approach-based analysis to be useful as a widely applicable tool, it must to some extent seem accessible and meaningful to other people than the author of this book. I am in no position to judge the extent to which this is the case.

A further challenge already discussed (in Chap. 2) is that representations in their published form rarely (if ever) contain exhaustive accounts of all the tacit assumptions, which are part of a particular representational practice. And not even the most rigorous analysis will be able to spell out all the propositional structures as well as conventional and skill-like practices involved in any given scientific approach.

As should be evident, then, I have no illusions that my discussions in this book will provide the final word in the debate regarding the epistemic virtues of interdisciplinarity. On the other hand, though, I do believe that my suggestions may contribute to moving these debates forward.

A Brief Reflexive Moment

I have argued above that no elements of representation are neutral. Neither tools, vehicles, assumptions, algorithms, observations, measurements, categorisations, nor theories. They all involve distortion to a certain extent. They are all perspectival, that is.

Is approach-based analysis itself neutral? Far from it, obviously! I have addressed this issue several times throughout the book. There is no doubt that approach-based analyses emphasise certain aspects while ignoring others and thereby represent targeted activities in somewhat distorted ways. Approach-based analysis is deliberately and consciously constructed in this way (as has been made quite clear above, I believe). It is the explicit aim of approach-based analysis to highlight issues hitherto neglected. This can only be achieved by paying less attention to the issues usually in central focus. Since I openly admit that social aspects of scientific activities are important in many ways, it is obvious that approach-based analysis, which bracket these aspects, cannot be exhaustive or undistorted.

From the perspective of approach-based analysis, interdisciplinary science is represented as combining elements from different parent approaches, whereas science in general is represented as an immense bundle (of bundles[1] (of bundles[2])) of approaches. Certainly, there are many other ways to represent science which might reach different but, perhaps, equally well-founded conclusions. Social frameworks, historical frameworks, logical frameworks, and so on, each contribute informative perspectives on science.

So even though I have argued that many of the social aspects on which conventional assessment of interdisciplinarity relies are more or less irrelevant to an epistemic assessment, I have also made it clear that these in many cases have a central role to play in explanations of why cases of integration end up having one particular structure rather than another. Science is certainly social, and social forces are very influential, of course—but they do not determine the epistemic vices and virtues of scientific representations.

In spite of the involved idealisations, I believe that approach-based analysis is a more viable tool for assessing epistemic aspects of interdisciplinary science than any existing approaches to the study of this phenomenon.

Future Opportunities

Perhaps a final conclusion may be that the above discussion has implications not just for the understanding of interdisciplinarity, but also for how to consider transdisciplinarity (in the sense of knowledge disseminating outside of academia). Even if we make the (idealised) assumption that what is disseminated are solely the outputs of the scientific activities, it does not seem unreasonable to assume that dynamics such as relocational idealisation are involved also when "knowledge" is imported and exported between scientific approaches and extra-scientific contexts.

Interesting future research possibilities therefore include the application of approach-based analysis to transdisciplinary cases. This would include tracking the impact of (increasingly idealised?) scientific knowledge as it moves through different sections of society. This should go hand in hand with further case studies of interdisciplinary science, of course.

Another interesting research question could focus on the extent to which using the framework of approach-based analysis during design phases might affect the development and outcomes of interdisciplinary projects.

Even though little of what has been discussed above is uncontroversial, and there is clearly a long way to go before matters of scientific evaluation are settled, it is my hope that the efforts invested in this book have resulted in an approach to the analysis of interdisciplinary research which might help evaluate concluded cases of interdisciplinary science as well as detect potential for fruitful adjustments in ongoing interdisciplinary activities. In this respect, it is worth emphasising once again, that explicit discussions of the ways in which interdisciplinary activities are supposed to result in scientific improvements are largely absent in existing treatments of the topic of interdisciplinarity. So just setting the stage for such discussions and providing some provisional means for carrying it out constitute considerable steps forward.

It is my hope (and belief actually) that the attentive reader will not leave this manuscript empty-handed, but will have original, more or less readily applicable tools at his or her disposal.

Notes

1. That is, disciplines.
2. That is, research projects.

Index[1]

---

[1] Note: Page numbers followed by 'n' refer to notes.

© The Author(s) 2018
R. Hvidtfeldt, *The Structure of Interdisciplinary Science*, New Directions
in the Philosophy of Science, https://doi.org/10.1007/978-3-319-90872-4

Printed in the United States
By Bookmasters